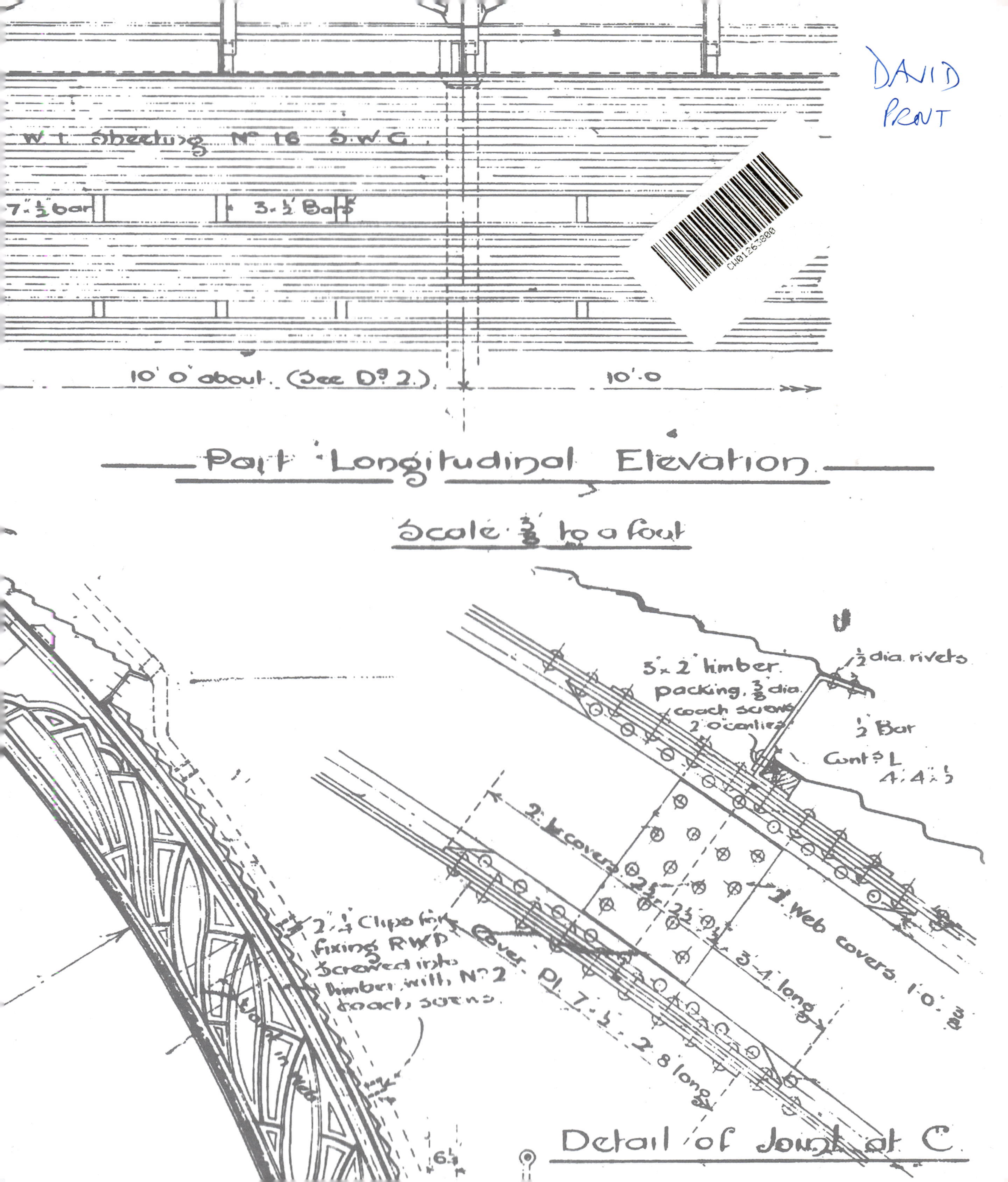
W.I. Sheeting No 16 S.W.G.
3 - ½" Bars
10' 0" about. (See Dg 2.)
10'.0
Part Longitudinal Elevation
Scale ⅜" to a foot
½ dia. rivets
5" x 2" timber
packing, ⅜ dia.
coach screws
2' 0" centres
½" Bar
Web covers 1'0"
Detail of Joint at C

CORRUGATED IRON

BUILDING ON THE FRONTIER

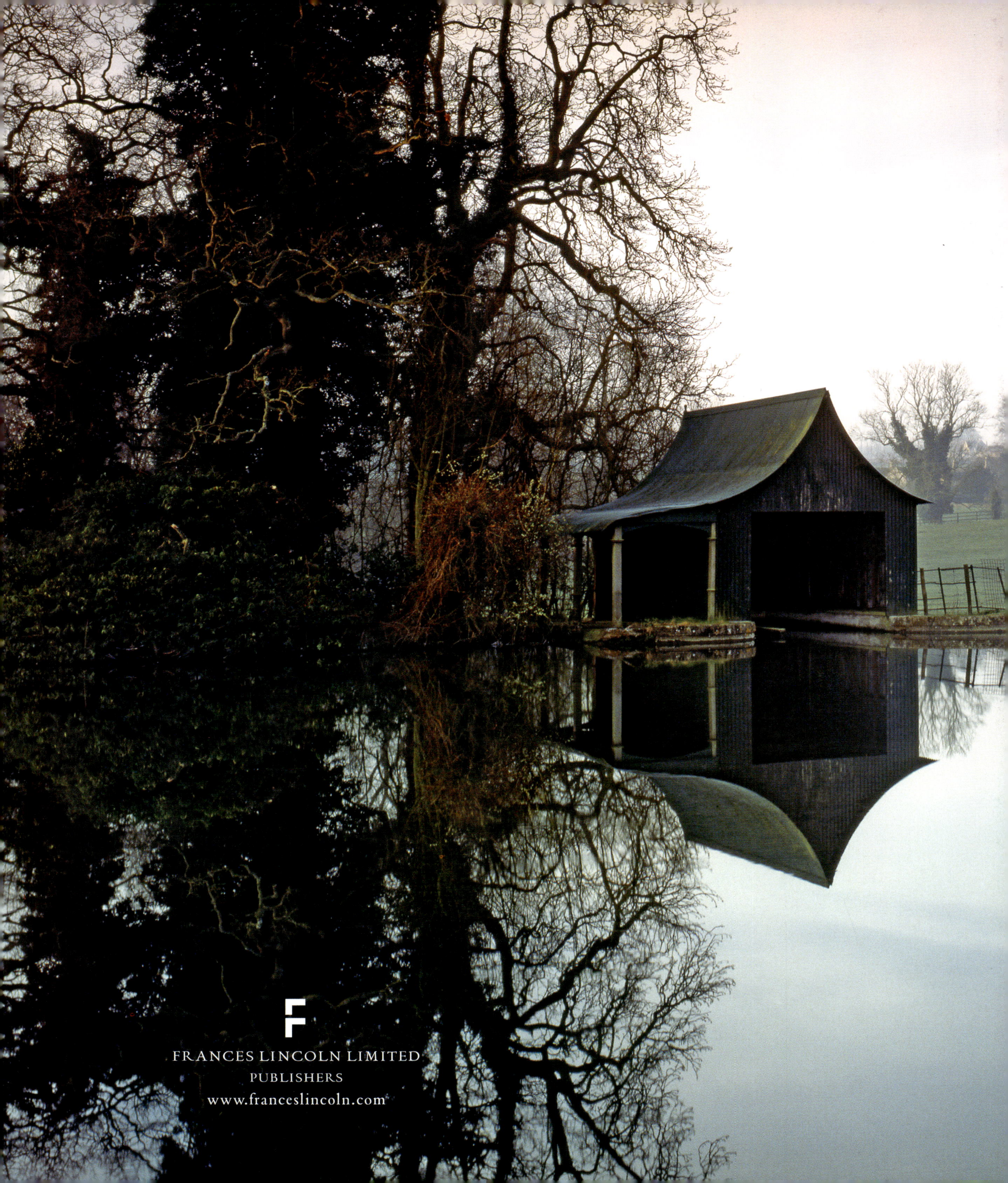
FRANCES LINCOLN LIMITED
PUBLISHERS
www.franceslincoln.com

Adam Mornement & Simon Holloway

CORRUGATED IRON

BUILDING ON THE FRONTIER

Frances Lincoln Limited
4 Torriano Mews
Torriano Avenue
London NW5 2RZ
www.franceslincoln.com

Corrugated Iron - Building on the Frontier

First Frances Lincoln edition: 2007

British Library Cataloguing in Publication Data. A catalogue record for this book is available from the British Library

ISBN 13: 978-0-7112-2654-8

Printed in Singapore

9 8 7 6 5 4 3 2 1

Commissioned and edited by Jane Crawley
Designed by Maria Charalambous

CONTENTS

Red Location, Port Elizabeth, South Africa. The name of the township was inspired by stains caused by rusting sheets of corrugated iron.

INTRODUCTION

Corrugated iron is a material of the frontier. It makes life possible in places that would otherwise be uninhabitable, whether due to extreme climate, inhospitable terrain, the scarcity of local building materials or the sheer scale of demand for shelter.

During the nineteenth century corrugated iron kit buildings provided secure accommodation for European colonisers seeking to establish a foothold in alien territories. Between 1916 and 1918 two and a half million British servicemen on the battlefields of Europe found refuge in corrugated iron barrack huts. Today it provides shelter for millions living in informal communities in the developing world. No other construction material is sufficiently affordable, transportable and strong to have addressed such a diversity of challenges.

But despite its many virtues, corrugated iron's contribution to society has rarely been acknowledged. One reason may be that there are no monuments made of corrugated iron, no grand buildings or landmarks. It is a humble, unpretentious material better suited to the background than the limelight. Another explanation is that corrugated iron has been a victim of its own success. Within a few years of its invention the material's virtues of versatility and affordability had stimulated its proliferation to the point of anonymity. It was everywhere, from shipyards to farmyards, tropical rainforests to Arctic wastes. This ubiquity created the impression that it had been around forever, but it had not. Corrugated iron was invented in 1829. Its claims to significance are abundant.

Once they have been corrugated (*ruga* is the Latin for wrinkle or wave) sheets of metal are extremely rigid. In this format sheets of iron became one of the few products of the Industrial Revolution to be absorbed into vernacular building repertoires and the first truly industrially-produced construction material to challenge the historic hegemony of timber, stone and brick.

The impact of the material on the public consciousness is evident in the survival of 'corrugated iron' as the generic term for all profiled metal sheeting. Iron has rarely been used since the early years of the twentieth century; corrugated iron's contemporary descendants are more likely to be steel, zinc, aluminium or any number of composites. Throughout this book 'corrugated iron' should be understood to refer to all forms of corrugated metal.

Thousands of civilians were displaced by unrest in the Ivory Coast during the mid-1990s.

Since the 1830s corrugated iron has been at the forefront of innovations in prefabricated construction, in large-span self-supporting roofs and in systems of mass-produced buildings. Around the world this lightweight, rigid material, employed in roofs and walls and countless other ways, has influenced patterns of farming, land settlement, light industry and warfare. It has helped to shape national and regional identities, notably in Australia and parts of America. In some Third World countries sheets of corrugated iron are among a person's most valuable assets and there are instances of corrugated iron forming part of a dowry and many examples of its use as a currency; in pre-genocide Rwanda, a prized Ankole cow could be traded for twenty sheets. Corrugated sheet metal may even be responsible for keeping the elements from more people's heads than any other material, a consequence of its prominence in the developing world.

Crucially, corrugated iron makes sense to people. It is recognisable everywhere and seems to invite practical involvement. But while an unskilled labourer can use recycled sheets to improvise a shed in an afternoon, gleaming new 'galvo' can also be used with precision in the design of finely crafted homes.

The material is also biodegradable, incombustible, earthquake resistant and thrives in diverse climates. A common view is that buildings made of corrugated iron are insufferably cold in winter and unbearably hot in warm climates, but that assumes a lack of refinement and is consistent with widespread negative associations with the material. Corrugated iron is only a barrier from the elements; it can be lined with layers of insulation appropriate to all types of climate. Since the early Victorian period buildings sheathed in corrugated iron have functioned effectively, and in some cases enduringly, in the steaming heat of Sri Lanka and the freezing cold of the South Atlantic.

OPPOSITE Lambing hut in Shropshire, England. Corrugated iron has been a feature of the British landscape for nearly 200 years.

LEFT Beach huts at Abersoch, North Wales. For older generations corrugated iron can trigger memories of seaside holidays or village hall dances.

Another reason why corrugated iron has been overlooked is that it has long been burdened by a perception problem. To many it is regarded as cheap, temporary and ugly; a crude material fit only for use in agriculture, industry or shanty towns. Social aspirants in the developed world have long been known to paint their metal roofs red, to evoke terracotta tiles, a superior and comparatively permanent material.

Depending on geographic and historic context, corrugated iron also carries political and cultural baggage. For instance, older generations in some parts of Japan associate corrugated sheet metal with a period of post-war deprivation and disgrace – parts of the war-damaged country were rebuilt with temporary metal buildings. But the material has not always been stained by negative connotations. In mid-nineteenth-century Britain sheets of galvanized corrugated iron were considered exciting and glamorous. Not only did they allow the construction of wide-span roofs without the need for heavy load-bearing walls, they also looked demonstrably new. In a world built predominantly of stone and timber, this reflective silvery material was compelling evidence of the pace of progress and the success of industry.

In 1838 the corrugated cast iron roof of the Coal Depot at the London Gas Works was described as, 'the lightest, most elegant and most economical roof yet [built]'. During the 1850s corrugated iron buildings erected prior to export – to check for defects and to number components – were regularly reported by the newspapers and became popular visitor attractions in their own right. On 13 May 1853 a church built in Bristol for export to Melbourne drew a congregation of over 800. The following year 25,000 people in ten days visited a Customs House built in Manchester for export to Peru.

Sadly, by the 1870s the novelty of corrugated iron was lost and it took public enthusiasm for the material with it. Nearly a century would pass before corrugated iron regained its allure, at least in parts of America and Western Europe.

The post-Second World War revival of enthusiasm for corrugated iron was driven by an interest in harnessing machine-produced building components in the development of new forms of architecture. Corrugated sheet metal, alongside other standardized industrial components, is a common denominator linking together some of the most revered figures of twentieth-century architecture and design, including Buckminster Fuller, Jean Prouvé, Charles and Ray Eames, Frank Gehry, Nicholas Grimshaw and Norman Foster.

Corrugated metal manufacturers still prosper, particularly in China, Australia, Turkey and several African countries. Throughout the developed world new sheets can be found in edge-of-town retail and industrial parks, in agricultural landscapes and in temporary hoardings to protect buildings at risk or under construction. Indeed, there are over a thousand different corrugated products on the market. As well as metals with different profiles and applications, it is possible to find corrugated plastic, cardboard, even bamboo in use as a building material. But few people mourn the demise of the surviving nineteenth and early-twentieth-century metal buildings, and even fewer appreciate their significance in the evolution of prefabricated buildings.

There are exceptions. To older generations corrugated iron is the trigger for happy memories of dances in village halls or holiday homes on the coast. To country dwellers sheets of rusting corrugated iron are a cherished feature of the landscape; others enjoy the reassuring sound of raindrops on metal roofs. Nevertheless, to the majority corrugated iron remains invisible.

This book was written to raise awareness of corrugated iron and its rich and varied history. It is a portrait of a wonderfully versatile and characterful material; one that for nearly two hundred years has been at the forefront of architectural and engineering innovations, which have in turn created shelter for millions, in diverse locations, at short notice and with a minimum of fanfare.

1 THE INVENTION OF CORRUGATED IRON

Most inventions are marginal alterations to existing products. Very few are as original or consequential as corrugated iron. But on 28 April 1829, when Henry Robinson Palmer, Architect and Engineer to the London Dock Company, registered Patent No. 5786, for 'indented or corrugated metallic sheets', he was simply responding to a need. He could not possibly have predicted the long-term repercussions of his actions.

Towards the end of the eighteenth century the major British ports, and London in particular, were choked with the success of their trade. Ships laden with imports or awaiting outgoing cargoes crowded the River Thames. The volume of traffic, compounded by the tides, meant that many suffered delays and great losses through theft and spoilage. The problem became the subject of parliamentary debates and numerous reports were drafted. The turning point came in the mid-1790s, when a committee of merchants supported a proposal to build a series of dock basins connected to the river via locks, allowing ships to unload at a quayside while not at the mercy of the tides. The new basins would also be built with large warehouses and high boundary walls, to reduce the threat of theft.

The London Dock, on the north bank of the Thames near the Tower of London, was one such complex. Excavation works began in 1800. Five years later the first phase of this enormous civil engineering effort was complete; a large rectangular basin surrounded by wide quays and five-storey buildings. It was one of the most advanced goods handling facilities in the world. The docks specialized in high-value luxury commodities such as ivory, spices, coffee and cocoa, as well as wine, turpentine and wool. There was space for 300 vessels and 200,000 tons of merchandise. But it was not big enough.

By the 1820s the London Dock had run out of capacity. A new basin and entrance locks were needed and in 1827 the London Dock Company appointed 32-year-old Henry Palmer to oversee their construction. He came highly recommended, having earlier worked for Thomas Telford, a pre-eminent civil engineer of the age.

The work involved excavation, moving large volumes of spoil and building dock walls. These were difficult operations. To remove the dug earth, Palmer harnessed steam power and drew on the talents of James Jones, his principal assistant, who started his career as a copper and tin worker but later became a specialist in the design and construction of machines. As well as the heavy

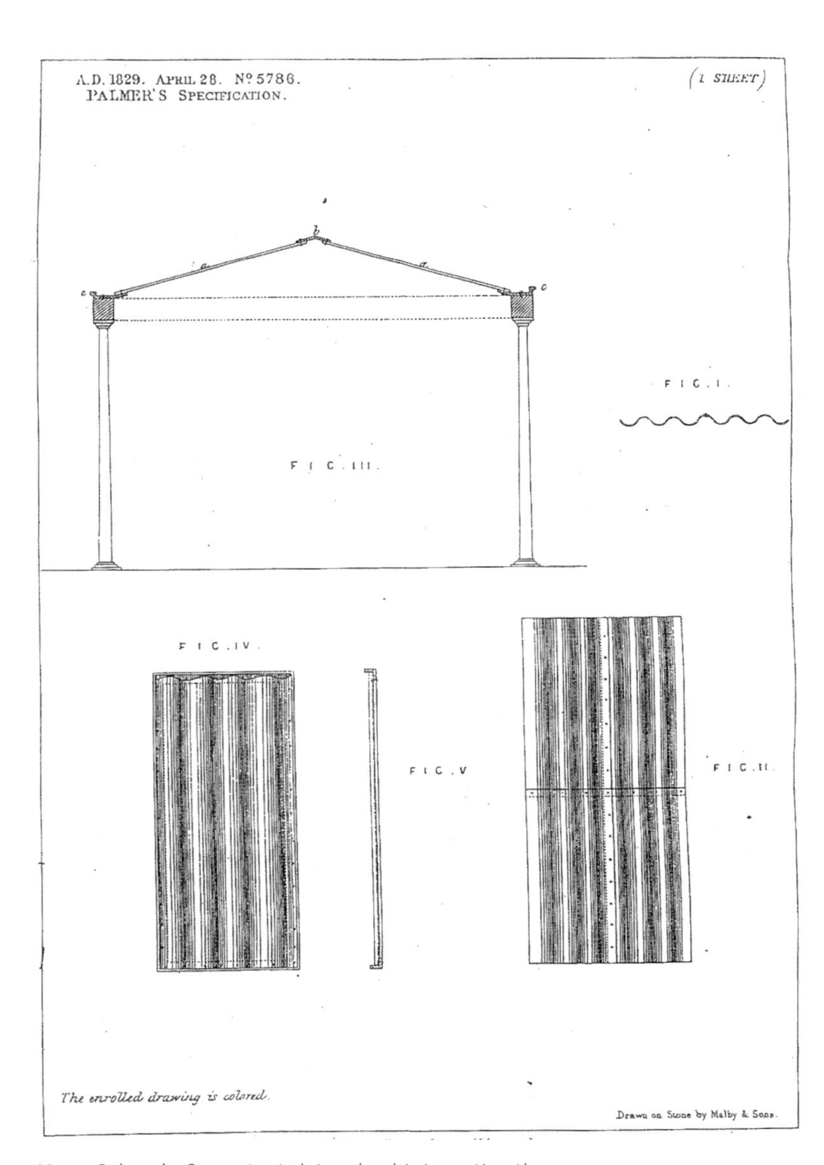

Henry Palmer's 1829 patent claimed as his invention the use of, 'corrugated metallic sheets… to the roofs and other parts of buildings'.

civil engineering work, Palmer was responsible for overseeing alterations and repairs to buildings in the 'old dock'. These were carried out with tenders on a job-by-job basis. From 20 November 1827 'Richard Walker, carpenter' often appears in the company's records. Over the next few years Walker and James Jones would play major roles in the evolution of corrugated iron.

Palmer's 1829 patent claimed as his invention: 'The use or application of fluted, indented or corrugated metallic sheets or plates to the roofs and other parts of buildings.' The drawing

The Turpentine Shed at the London Docks, c.1830, was the world's first building with a corrugated iron roof.

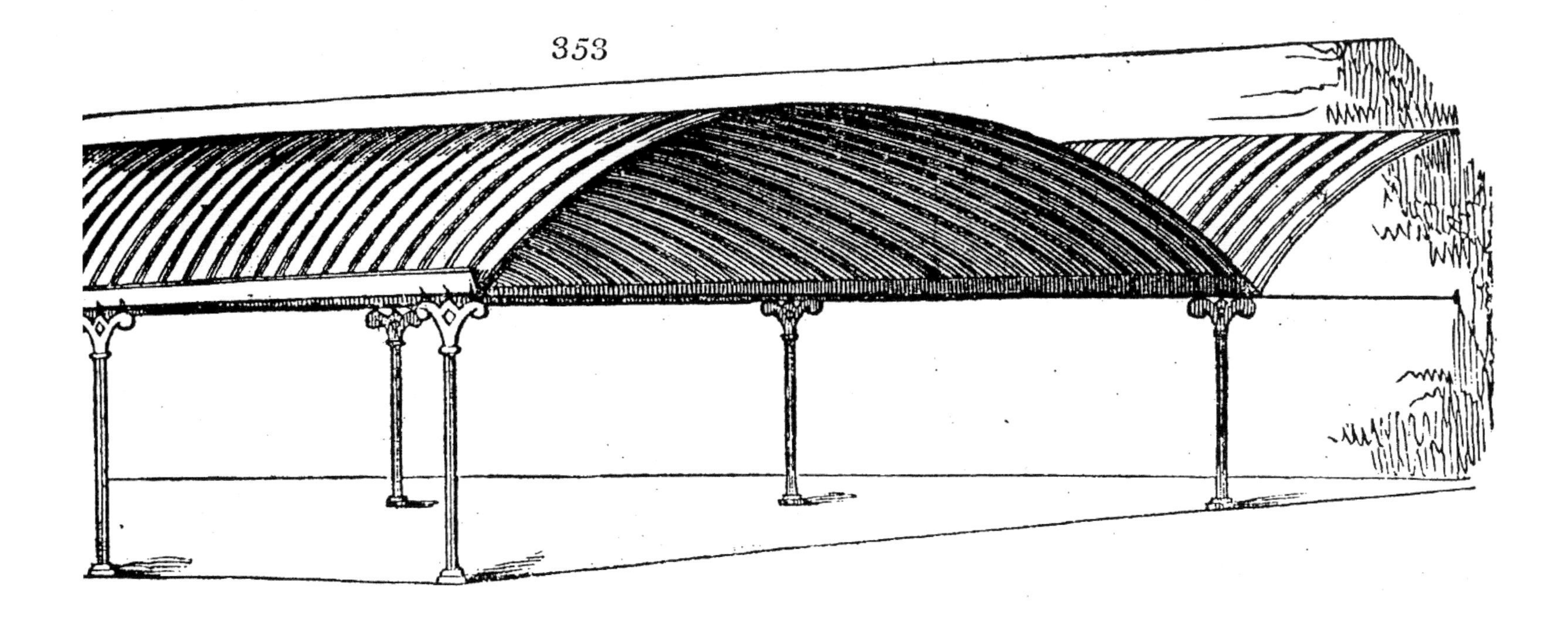

shows a modest pitched roof without framing, emphasizing the self-supporting rigid properties of his revolutionary lightweight panel, which could also be used for walls in modular units or as continuous overlapping sheeting. But Palmer was in a quandary. He was not in a position to capitalize fully on his invention.

Palmer was an engineer, a man of ideas. He could solve intellectual problems and he could identify people with skills and expertise relevant to his current concerns. But he was also extremely busy. His work at the docks left him little time to get to grips with the challenges that manufacturing corrugated iron would entail. This may have been why he sold his patent to Richard Walker a few months after it was sealed.

The world's first corrugated iron building

In June 1829 Palmer, 'laid before the [dock] committee a plan for the erection of a shed on the east side of the new dock'. The following January the committee agreed that, 'Mr Palmer be instructed to prepare a plan and estimate for the expense of erecting a shed with a patent iron roof to cover the said vaults'. The Dock Company purchased the flat sheet iron by tender from agents in London. The process of corrugating and curving the iron took place at or near the dock, perhaps using the steam engine that had been used in the excavations.

In those early days the process of corrugating involved passing iron sheets through fluted rollers – the latter may have been heated. This method limits the maximum length of sheet to the overall width of the rollers, but it does not demand a great deal of power and does not stretch the sheet metal. Over the years other approaches were tried, including stamping corrugations one-by-one and forming iron tiles using dies. But the use of fluted rollers was the easiest and most obvious way of manufacturing corrugated iron using machinery, and it remains commonplace today.

Many payments to Richard Walker are recorded for work specifically connected with corrugated iron at various times and places in the dock. The earliest mention is in January 1830, when Walker was paid for, 'erecting 22 additional squares of patent iron roofing to the Turpentine Shed on the west side of the entrance basin'. (A square in this context is 100 square feet, about 9 square metres). This building was probably the prototype on which the idea was tested prior to its more extensive use elsewhere at the dock. It was also the building illustrated in John Loudon's

BELOW Section through a rolling machine designed in 1844. Palmer, Walker and Jones would almost certainly have used a similar device at the London Docks.

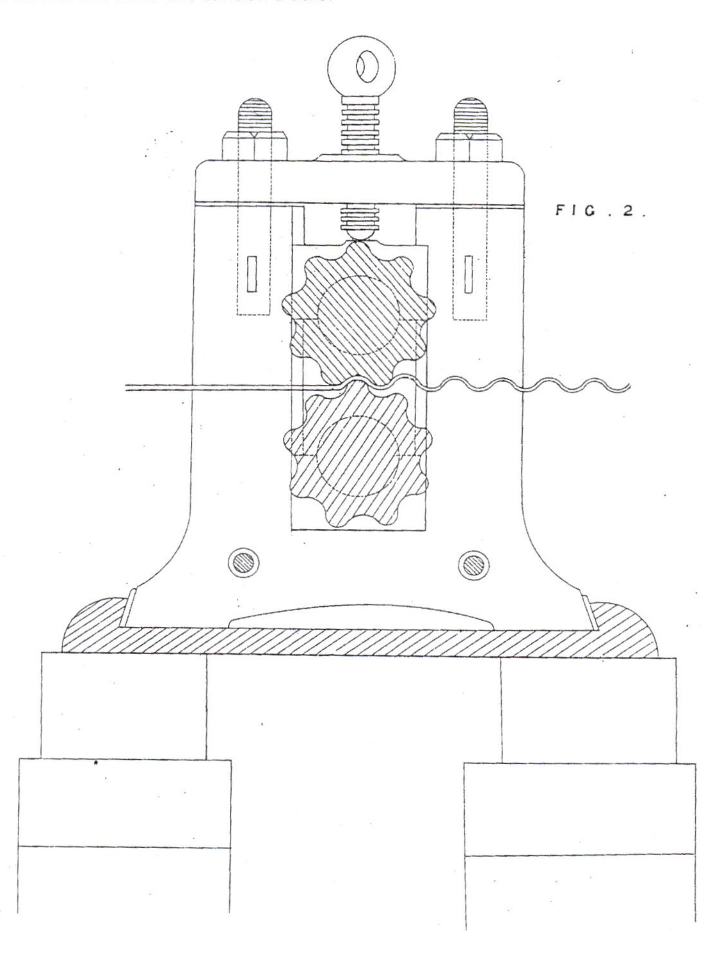

An Encyclopaedia of Cottage, Farm and Villa Architecture and Furniture, a popular architectural sourcebook of the time, and almost certainly the building described by an anonymous contemporary writer in the *Register of the Arts and Sciences*:

> *Extraordinary, light and simple roof*
> On passing through the London Docks a short time ago, we were much gratified in meeting with a practical application of Mr Palmer's newly invented roofing. This singular roof, supported by light cast-iron pillars, forms a shed on one side of the basin near Wapping Church, and covers an area of about 4,000 feet [the discrepancy with the Minute Book reference may be explained by journalistic hyperbole]. Every observing person, on passing by it, cannot fail being struck (considering it as a shed) with its elegance and simplicity, and a little reflection will, we think, convince them of its effectiveness and economy. It is, we should think, the lightest and strongest roof (for its weight), that has been constructed by man, since the days of Adam. The total thickness of this said roof, appeared to us from a close inspection (and we climbed over sundry casks of sticky turpentine for that purpose,) to be, certainly not more, than a tenth of an inch! although stretching over a space of about 18 feet, by a slightly curved arch. It is composed wholly of malleable-iron plates or sheets, about two feet wide and five feet long; and each plate is so bent, as to form a series of uniform undulations, producing thereby, alternate grooves and ridges longitudinally which serve as water courses for the

LEFT A painting of a Lysaght & Co. plant in New South Wales showing employees corrugating galvanized iron with fluted rollers, *c.*1940s.

RIGHT Page from a Richard Walker advertisement published in 1839.

78 Appendix.

By the King's Royal Letters Patent.

RICHARD WALKER,

Carpenter & Builder,

Manufacturer of the PATENT CORRUGATED IRON

ROOFS, DOORS, SHUTTERS, PARTITIONS, SAFETY ROOMS, PARK ENCLOSURES, VERANDAS,

AND ALL KINDS OF

PORTABLE BUILDINGS FOR EXPORTATION,

Fort Place, Grange Road, Bermondsey,

LONDON,

AND 6, GARDEN ROW, CAMBERWELL.

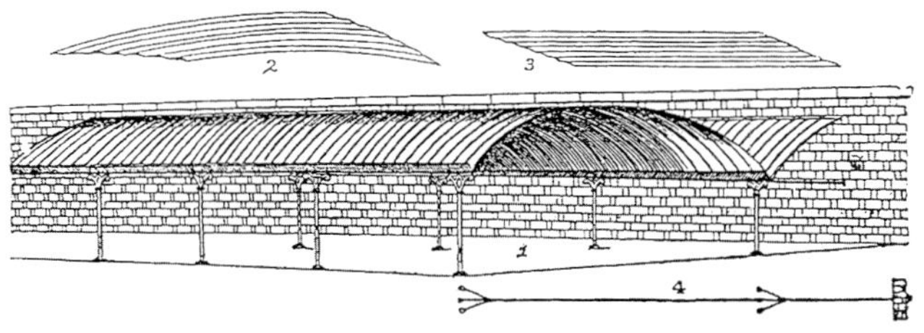

A new property is given to Sheet Iron by its being Corrugated, or formed, by means of powerful Machinery, as continued Arches, as shewn in Fig. 3 of the above Diagram. A sheet of Iron so thin that it will not sustain its own weight, will, after this process, bear 700 lbs. It sustains no injury from the weather when covered with a slight coat of paint; effects a great saving in other materials, as Buildings may be entirely constructed of it; and it is capable of being applied to the most ornamental as well as useful purposes.

The Corrugated Iron may be seen, on an extensive scale, in the London Docks. The Drawing, Fig. 1, represents a Shed in the Entrance Basin, the length of which is 225 feet, its width 40 feet, and the height of the columns on which the roof is supported, 12 feet. The Columns are of cast iron, a Gutter of which metal is continued from column to column, the whole length of the roof; and in the gutter rests the edge of the roof. The arch is formed of several sheets of iron curved in the transverse direction to the corrugated arches, as described, Fig. 2, and riveted together longitudinally: a continuation of similar arches connected to each other by rivets, compose the roof: every corrugated arch forms a watercourse ending in the gutters at the side, thereby rendering it quite waterproof. To give stability to the whole, a Tie Rod, Fig 4, is carried across the shed from each column to its opposite. Between this shed and the brick-wall is a Leanto Corrugated Roof, forming a half arch, springing from the gutter, and resting against the wall, as represented on the right of the sketch. There are also several Sheds, Gates, and Enclosures in the Eastern or New Dock, on the other side of Old Gravel lane, and at the Hambro Wharf, Thames street, near Southwark Bridge; at the Patentee's Manufactory, in the Grange Road, it is applied in various ways which may serve for the purpose of inspection.

It is particularly recommended for PORTABLE BUILDINGS FOR EXPORTATION, The small space occupied in stowing them, when the respective parts are separated, rendering their conveyance cheap and easy. For New Settlements, the facility with which they may be erected, or removed from place to place, is a desideratum of great consideration.

Barns, Granaries, and other Agricultural Buildings may be constructed of it, relieving the Farmer from all dread of Nocturnal Incendiaries, who have committed such alarming ravages in many parts of the country. It is used in the formation of Doors and Window Shutters to Houses of brick or other materials, where particular security is desirable, its great hardness preventing its being cut through, also in the erection of Strong Rooms, Safes, Chests, Fire-proof Buildings, and Apartments used for the preservation of Records and Valuables: a double lining, with an air space between, being an effective barrier to the greatest intensity of flame.

rain. This grooving, or as we might say, arching and counter-arching, confers great strength to resist the longitudinal and vertical pressure to which roofs are subjected. Every plate appears also to be slightly arched in the transverse direction, and as they afterwards form a portion of the curved line of the roof, a shell-like stiffness is given to the plates, which enables them to resist a greater force than they would be able with plane surfaces, were they several times their present thickness. It is also worthy of observation, that the inconvenience usually resulting from the expansion and contraction of such extended surfaces of metal by variations of temperature, is, by Mr Palmer's arrangement of the grooves and arches in opposite directions, in a great measure, if not entirely, obviated.

The sight of the material was clearly a shock to contemporary eyes. The convoluted description also suggests that the term 'corrugated iron' was not yet commonly used.

Self-supporting roofs

Palmer's 1829 patent envisaged two principal ways for corrugated sheet iron to be used: as cladding on a supporting framework, and as self-supporting sheets capable of spanning across space without any framing. The latter, the more revolutionary method, was used in the shed at the London Dock. It was a system that gave the buildings a distinctive barrel-shaped roof, a form that has since come to define the material.

The system used at the dock consisted of parallel rows of cast iron pillars which were given rigidity by cast iron guttering running the length of the structure, and tie rods running between each pair of pillars. Sheets of corrugated iron were curved and riveted together to form a self-supporting arch. These roofs depend entirely on the rigidity of the corrugated sheet iron to achieve the required span. Tie rods prevent the roofs spreading out.

Up to the late-1840s this method was used on a number of large-span enclosures. Notable examples include the Eastern Counties Railway station in Shoreditch, East London, and the London Gaslight Company's Nine Elms gasworks in Battersea, south of the Thames. Unsurprisingly, given that he held the exclusive rights to the use of the material, Richard Walker both supplied the corrugated iron and built these two structures.

The Eastern Counties Station, on which work began in 1839, was the London terminus of the Northern and Eastern Railways. It was designed by John Braithwaite, engineer-in-chief to the Eastern Counties Railway and an early pioneer of steam-powered railway engines. The station cover consisted of three elliptical roofs of corrugated iron supported by structural cast-iron gutters on two rows of 17 cast-iron columns and the brick walls of the station. The central span was 36 feet wide; the roofs of the two side buildings spanned 20 feet 6 inches, rising 4 feet at the centre. The station was 230 feet long, enclosed at the west end. It covered over 17,000 square feet (nearly 1,600 square metres), an enormous area for a roof without any conventional roof framing in the form of trusses, purlins and rafters.

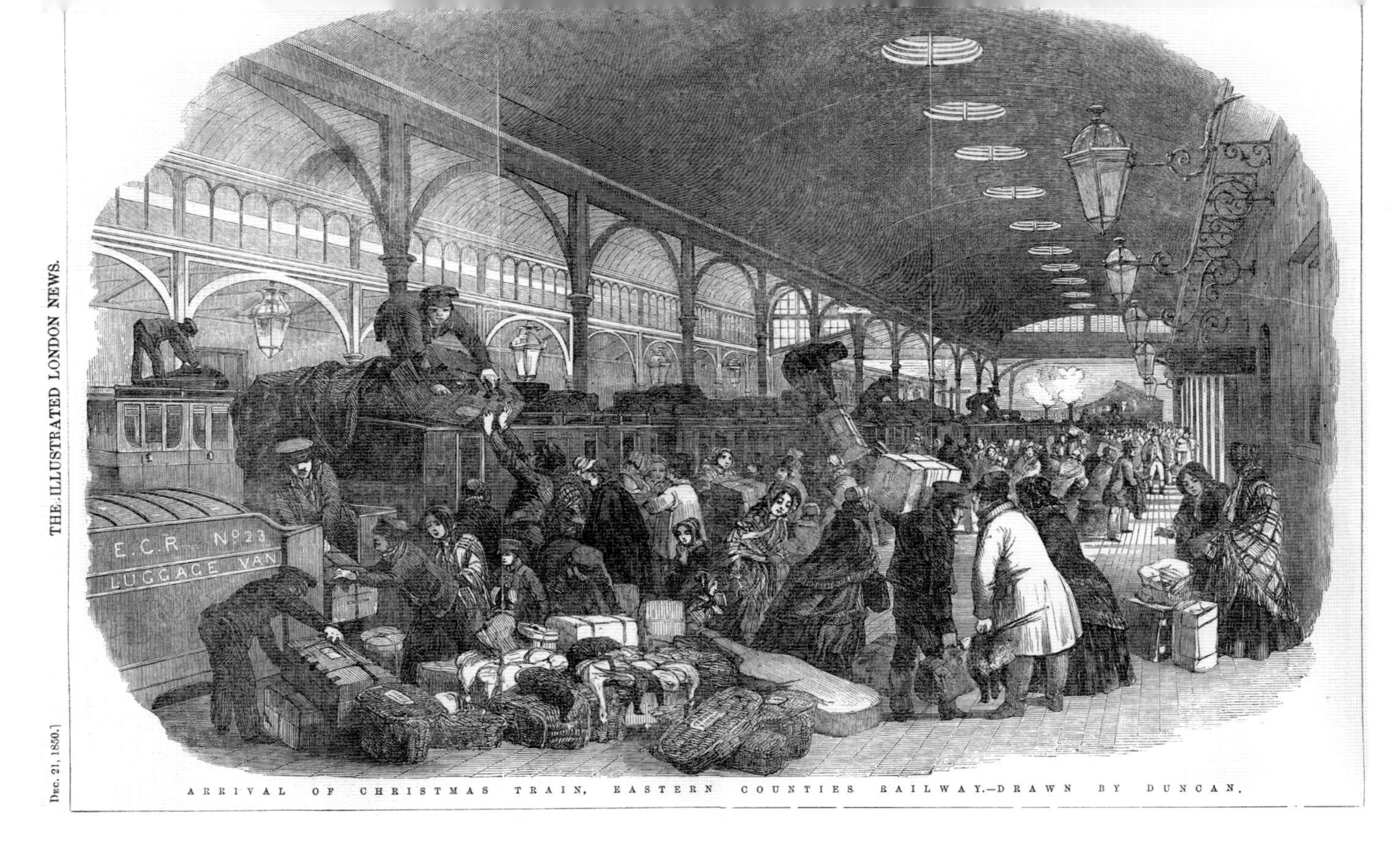

Uncovered holes allowed light through the roofs of the earliest corrugated iron structures, such as the London terminus of the Eastern Counties Railway.

In the late 1840s corrugated iron covering an area of three acres was supplied by Richard Walker to Price's Candle Works in south London.

'Semicircular roofs of corrugated iron supported on light iron columns', enclosed an extension to Billingsgate New Market, c.1850s.

From the late-1840s and on into the 1850s, Walker supplied the iron for a number of factories, including several for Price's Candle Works. The first, which had an iron roof covering three acres, was erected at Battersea, before it was relocated to the Wirral, near Liverpool.

Another example of a self-supporting roof was Billingsgate New Market. In the early 1850s, London's wholesale fish trade was booming, so City architect James Bunning was charged with enlarging the existing market premises. This he achieved with the addition of, 'semicircular roofs of galvanised corrugated iron, supported on light iron columns and girders, with sky-lights facing the north ... a glass roof also runs from north to south adjoining the houses on the west side of the Market'.

These early experiments with self-supporting roofs coincided with parallel experiments in the protection of sheet iron from the elements. A weakness of corrugated iron was its propensity to corrode. Although Walker claimed that, 'it sustains no injury from the weather when covered with a slight coat of paint,' the salt laden air of the tidal Thames at the London Docks and the corrosive atmospheres inside coal and gas retort houses and, most publicly, railway stations and sheds, caused the sheets to deteriorate rapidly.

The first patent to address the problems was filed by Stanislaus Sorel, a civil engineer working in Paris, in May 1837 in France. This coincided almost exactly with a British patent (No. 7355) filed by Commander H. V. Craufurd, RN. It is possible that Craufurd was Sorel's agent in Britain. Both claimed as novel a process of coating iron with a thin coat of zinc to render the iron resistant to corrosion. Craufurd, in particular, mentioned that the zinc coat would be suitable for protecting sheets of iron for roofing and other purposes. The potential of this process – now known as hot dip galvanizing – was clear to a number of engineers and manufacturers. But in the 1830s, and for at least

Gloucester Docks, West England (now demolished).
Self-supporting arches bound by tie rods quickly became a signature of corrugated iron roofs.

the next thirty years, the chemistry was not fully understood.

As the years passed, on-the-ground experience and practical innovations improved the reliability of the end product. The early 1840s saw a rash of activity. First Craufurd's patent was challenged and found to have a precedent, forcing him to withdraw some of his claims. This was followed by countless other patents claiming improvements and modifications. One of them was filed by Edmund Morewood, a London merchant. His patent of 1841 (No. 9055) involved coating the iron with tin before dipping the whole into zinc to give a much shinier appearance, a finish that proved particularly popular in the United States.

Arched roofs without any supporting framework, apart from tie rods, continued to be used throughout the nineteenth century. Indeed, during this era of experimentation, immense feats of engineering were achieved. In 1847 John Porter, one of the great early innovators with the new material, designed a self-supporting roof spanning an incredible 46 feet (14 metres) for the Patent Fuel Company in Swansea. But the popularity of this revolutionary roof type was tempered by suspicion. Writing in 1868, contemporary commentator Henry N. Maynard summed up the mood:

> This is an inexpensive kind of roof and in spans not exceeding 35 feet may be employed with advantage for sheds or buildings... It is however, necessary to say that the amount of material of which they are composed is quite inconsistent with the result of calculation upon which the details are designed... [However] they do often withstand... the periodic gales in a manner difficult to reconcile with the small amount of material in them.

The problem was that all the self-supporting structures erected during the nineteenth century were the product of trial and error; it was impossible to subject them to rigorous mathematical analysis. By the beginning of the 1860s this uncertainty rendered them inappropriate for use in large-scale public buildings. Instead they found favour in ephemeral structures, notably farm buildings.

Overlapping sheets

The other chief use for corrugated iron envisaged in Palmer's 1829 patent was for overlapping sheets to form a cladding for walls, or covering for roofs, supported on a frame of timber or iron. By the early 1840s this had become the most common use of the material, and it remains so today.

Some of the earliest applications of the material in this form were for large-span Navy slips and railway termini. Its popularity in these buildings is explained by the fact that during the 1830s and 1840s both the Navy and the railway companies needed large-span roofs unhindered by intermediary supports to meet their respective requirements, and corrugated iron fitted the bill perfectly. The sheets were lightweight, compared with other roofing options, and boarding and battens became surplus to requirements because they could be used to span large distances. Now the roofs could be made with trusses and purlins with the sheeting fixed to them. The reduced fire risk was another major consideration.

Throughout history ships had been built in the open air, exposed to the same elements that the vessel would experience during its service. However, alarming experience gained during the Napoleonic Wars (1799–1815) convinced the Navy Board that ships and their builders would fare better if housed in dry conditions during construction. A wooden vessel might take several years to complete and carefully seasoned timber would begin to degrade when left open to the elements during this time. This was significantly shortening the vessels' lifetime; dry rot was setting in during the period of construction. The argument for building ships under cover

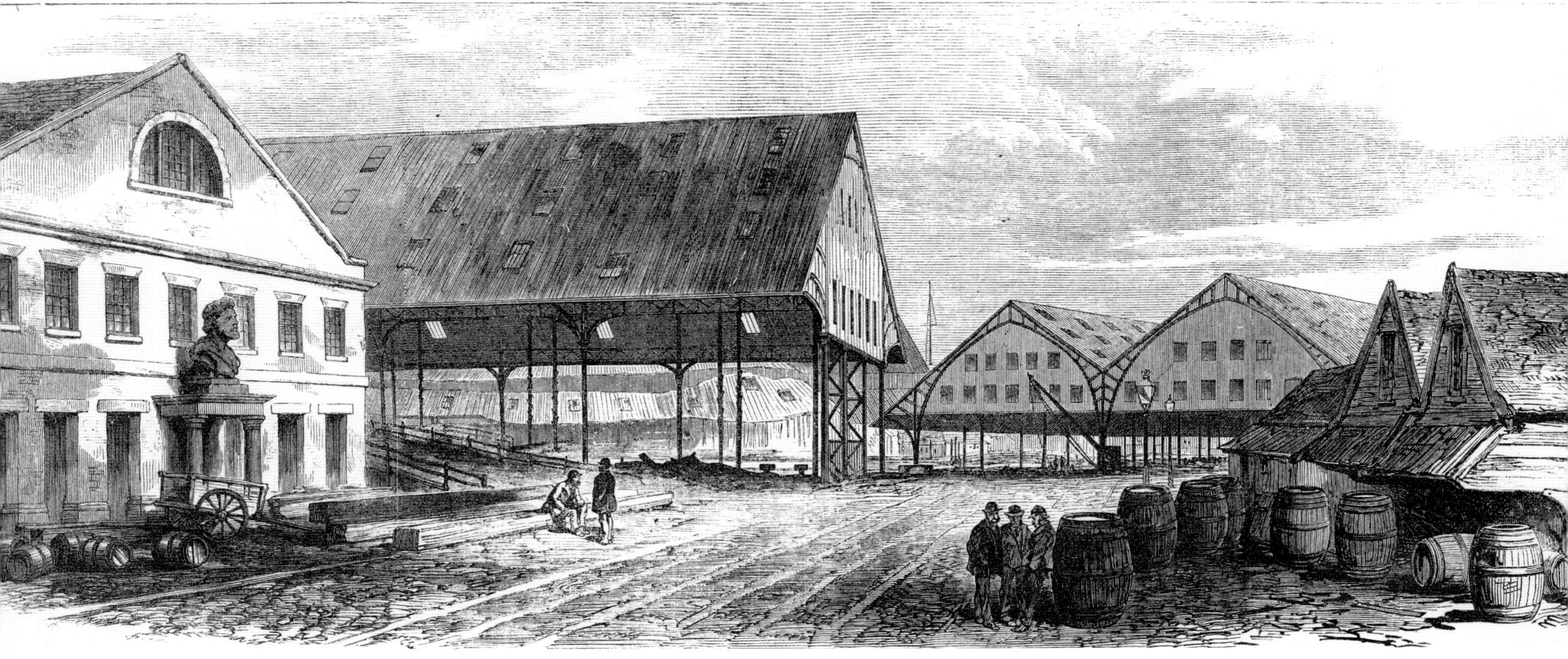

The three slips at Deptford Docks, London were built during 1845 and 1846 by George Baker & Son.

was further strengthened by the potential of losing less construction time to inclement weather and being able to extend working hours with the aid of gas lighting. But the challenge was to create slips large enough to build the new fleet. This required radical thinking.

Up to the 1830s slips were made of timber. But the material had been stretched to its limit. Although adequate spans could be created, space in the upper parts of the buildings – where the slope of the roof begins – was becoming increasingly restricted, particularly for the new breed of vessel that the British Navy had in mind to counteract the French threat. A new phase of slip building began in 1837, when Captain Henry Brandereth R.E. was appointed Director of Naval Works. For at least the next twenty years Royal Engineer officers were prominent in the design of buildings and civil engineering works at naval shore stations.

The first contract to supply iron roofs for the Navy was won in 1844 by Fox, Henderson & Co. of London Works, Birmingham. The company was challenged to use corrugated iron sheeting to replace the severely decayed roofs over slips 8 and 9 at Pembroke Docks in Wales. Construction of the £15,480 tender began in September of that year. The two roofs were identical – 262 feet long and 58 feet wide. Each corrugated iron roof was supported on 48 cast iron columns joined by girders and trusses. Numerous tie rods, struts and wind braces were included in the structure and, already riveted together in groups of four, the galvanized corrugated sheets were hauled into position and workmen on wooden seats slung from the superstructure bolted them to the purlins.

In all, sixteen slip roofs framed in cast and wrought iron and clad in galvanized corrugated iron were built between 1845 and 1847. They enclosed the major slips at Pembroke, Woolwich, Portsmouth, Deptford and Chatham naval yards (see pages 21–23). Five were the work of Fox Henderson; George Baker & Son built all the others. Two other iron roofs, notable for the incorporation of 'hanging railways' (overhead cranes), were completed at Chatham and Woolwich in 1855. Colonel Godfrey Greene, who had trained with the Royal Engineers at Chatham before his appointment as Director of Engineering and Architectural Works at the Admiralty, designed both of them.

And then the age of slip building roofs ended. In part this was brought about by the introduction of iron-hulled ships, although a more significant factor was that the new vessels were too large for even the most ambitious Victorian to consider protecting under cover. But the significance of the iron-roofed slips in architectural history should not be underestimated. Indeed, the roof of the 1851 Great Exhibition building might not have been possible without them.

When doubts were expressed over the stability of the Crystal Palace roofs, Charles Fox of contractor Fox Henderson said: 'That was what they said about our Pembroke Dockyard building sheds; they were some 100 feet high [an exaggeration in a good cause] and more than that wide and the shipwrights refused to work in them for fear of their slim and delicate outline giving way.'

One final footnote to the story of corrugated iron and the Navy, which had been forgotten about until it was rediscovered in the late 1950s, was the groundbreaking structure now known as the

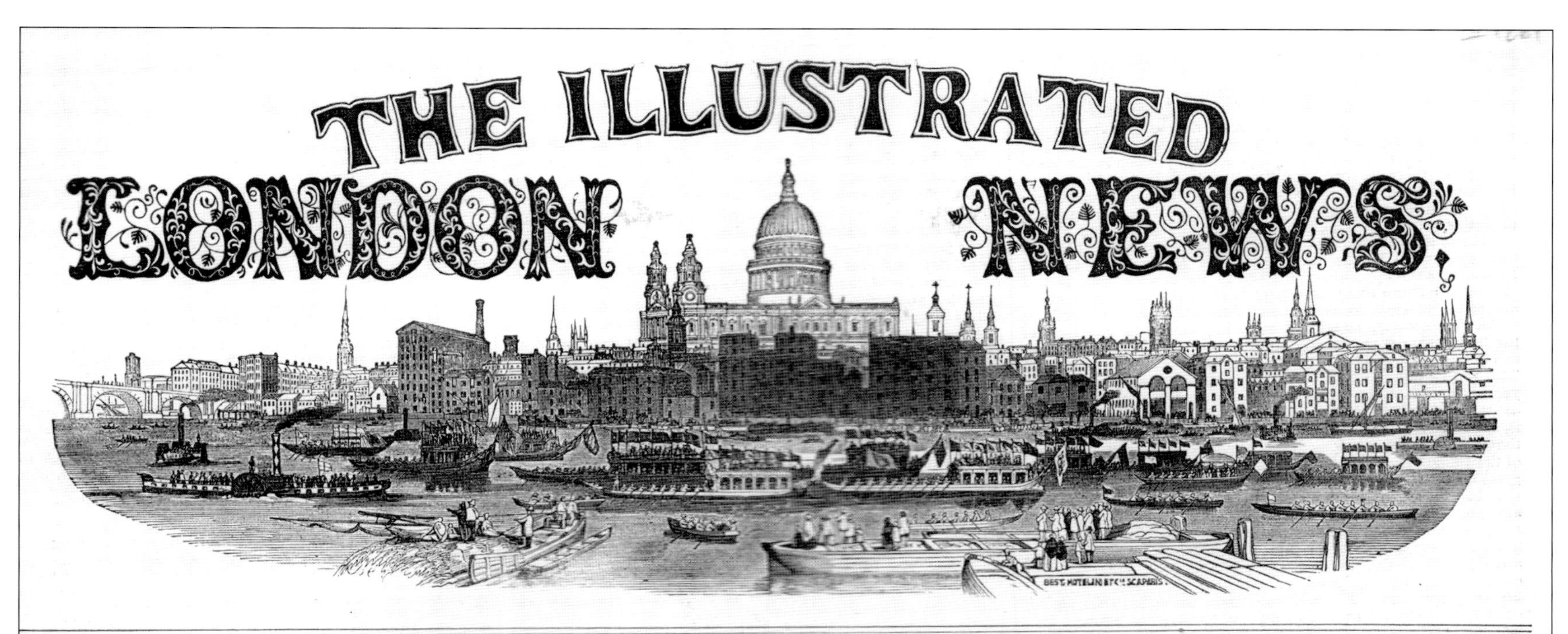

No. 619.—VOL. XXII.] FOR THE WEEK ENDING SATURDAY, APRIL 30, 1853. [SIXPENCE.

LAUNCH OF HER MAJESTY'S SCREW STEAM-SHIP-OF-WAR, "JAMES WATT" (90 GUNS), AT THE ROYAL DOCKYARD, PEMBROKE.—(SEE NEXT PAGE.)

Sheerness Boat Store (built 1858-60), is a very early example of a multi-storey iron-framed building and one of the most revolutionary buildings designed by Colonel Godfrey Greene.

Boat Store in H. M. Dockyard at Sheerness. The Boat Store – another of Godfrey Greene's progressive works – was built between 1858 and 1860, but it appears to belong to a different age entirely. Great claims can be made for the structure. It is a very early example of a multi-storey iron-framed and corrugated iron clad building, and the first to use H-section columns and beams rather than the elaborate and decorated forms typical of the period. It is also considered a well-proportioned design with sensitive use of materials, pre-empting many ideas that gained currency in modern architecture over a century later.

Railway termini

The other main clients for large-span roofs composed of overlapping sheets of corrugated iron were the railway companies. The myriad benefits of the material for the railway companies compared with conventional roofing systems, particularly in their vast showpiece termini, were described by Richard Turner of the Hammersmith Iron Works in Dublin, the company that supplied the roof for Liverpool Lime Street Station.

> It may be well to consider some of the prominent objections to the ordinary railway roofs, so as to establish reasons for the adoption of the system [iron roofs] now proposed. Many of the roofs, now in general use for railway stations, have... more wood than iron in their construction, and they are, moreover, covered with slates. Now, a timber roof is subject to decay... and is also liable to be consumed by fire, thereby endangering the property it is intended to cover, and, when placed in connexion with iron, it is an uncertain material, though the iron truss is frequently made to rely upon it for stability. Again, slates, as a covering, are bad, because strong winds will draw the nails intended to secure them; and being consequently blown away, the wood is left exposed to the weather. These disadvantages are, however, slight, when, compared with the objectionable form of high pitch, and consequent depth of trussing necessary for roofs covered with slates; when, therefore, a large area is to be covered, the space must be divided into a number of widths, or spans, forming a complete series of roofs, carried by a dangerous and inconvenient range of supports, or columns, which, in a railway station, are always liable to be knocked down by any engine, or carriage, that may, by accident, get off the line: thus placing the whole structure, as well as the lives of the passengers and servants, in imminent danger.

By the later decades of the nineteenth century, corrugated iron and railways were inseparable. Aside from the glamorous terminals, corrugated iron was used in the construction of countless maintenance and storage buildings, and in the halts and stations on the smaller branch lines. The intensity of the relationship is perhaps no surprise: the growth of the railways coincided with the development of corrugated iron. But the origins of the relationship were tempestuous. Several of the rail companies needed plenty of encouragement to acknowledge the benefits of the new material.

The directors of the London and North-Western Railway Company were not at first convinced by the proposal that a single arched roof at Lime Street could be built and would be practical. They demanded that Richard Turner conduct rigorous tests on a full-size portion of the proposed roof at his works, and at his own risk. The experiment was successful and the largest span roof built

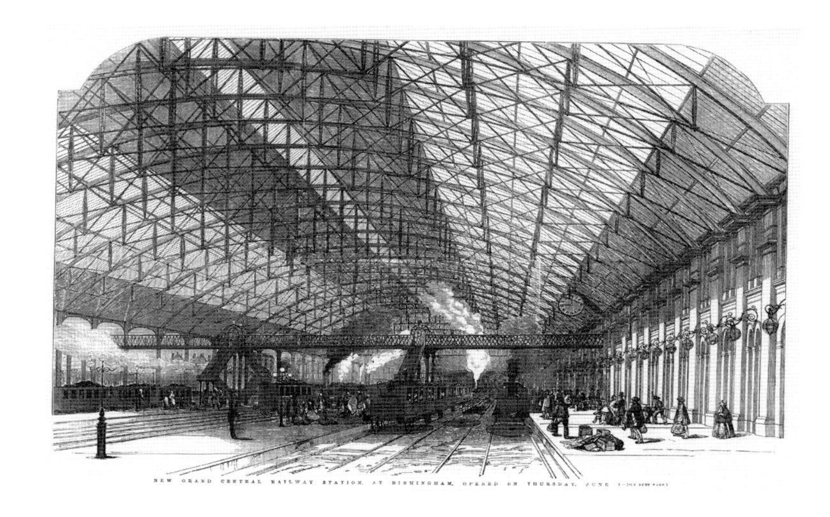

RIGHT The 212-foot roof span of Birmingham's 'New Grand Central Railway Station', officially opened on 1 June 1854, was, 'the largest hitherto attempted'.

BELOW The builder of Liverpool Lime Street station was required to erect and test a full-size section of the roof before the directors of the North-Western Railway Company granted him the building contract.

to that date opened in 1850. It was fixed along one edge to the brick station building and, 153 feet 6 inches (47 metres) away, onto 17 cast iron pillars. It was 374 feet long, cost £15,000 and took ten months to construct.

Four years later Birmingham New Street Station superseded Lime Street. Built as the Joint Station for five rail companies, New Street was necessarily large. The commission was further complicated by the requirement that construction took place while passengers and trains moved underneath. For a year Messrs Fox, Henderson and Company, following a design by E. A. Cowper, constructed the latest 'largest span hitherto attempted' of 212 feet (65 metres). It was shoehorned into a very tight, irregular space in the centre of the city, and was 840 feet long. Bowstring trusses of innovative design were used in its construction, and 64,000 square feet of galvanized corrugated iron sheets covered half of the roof area; the remainder was made of fluted plate glass. The total cost was £32,274.

The other landmark station to open in 1854 was Paddington (see pages 24–27) in West London by Isambard Kingdom Brunel, another early innovator in the use of corrugated iron. Typically, Brunel used the sheets in an unconventional manner, laying the corrugations at right angles to the slope of the roof, for additional strength and to eliminate one layer of framing.

Primed for take-off

The years following Palmer's patent were a time of trial and error. The early innovators thought on their feet to find the best ways of optimizing the characteristics of this new material. Along this previously untravelled path they experimented with techniques for creating corrugations, protecting the material from corrosion in the elements, applying it to large-span frames and using its inherent rigidity to create self-supporting structures.

As the nineteenth century reached its halfway point, this process of experimentation was by no means over. There were

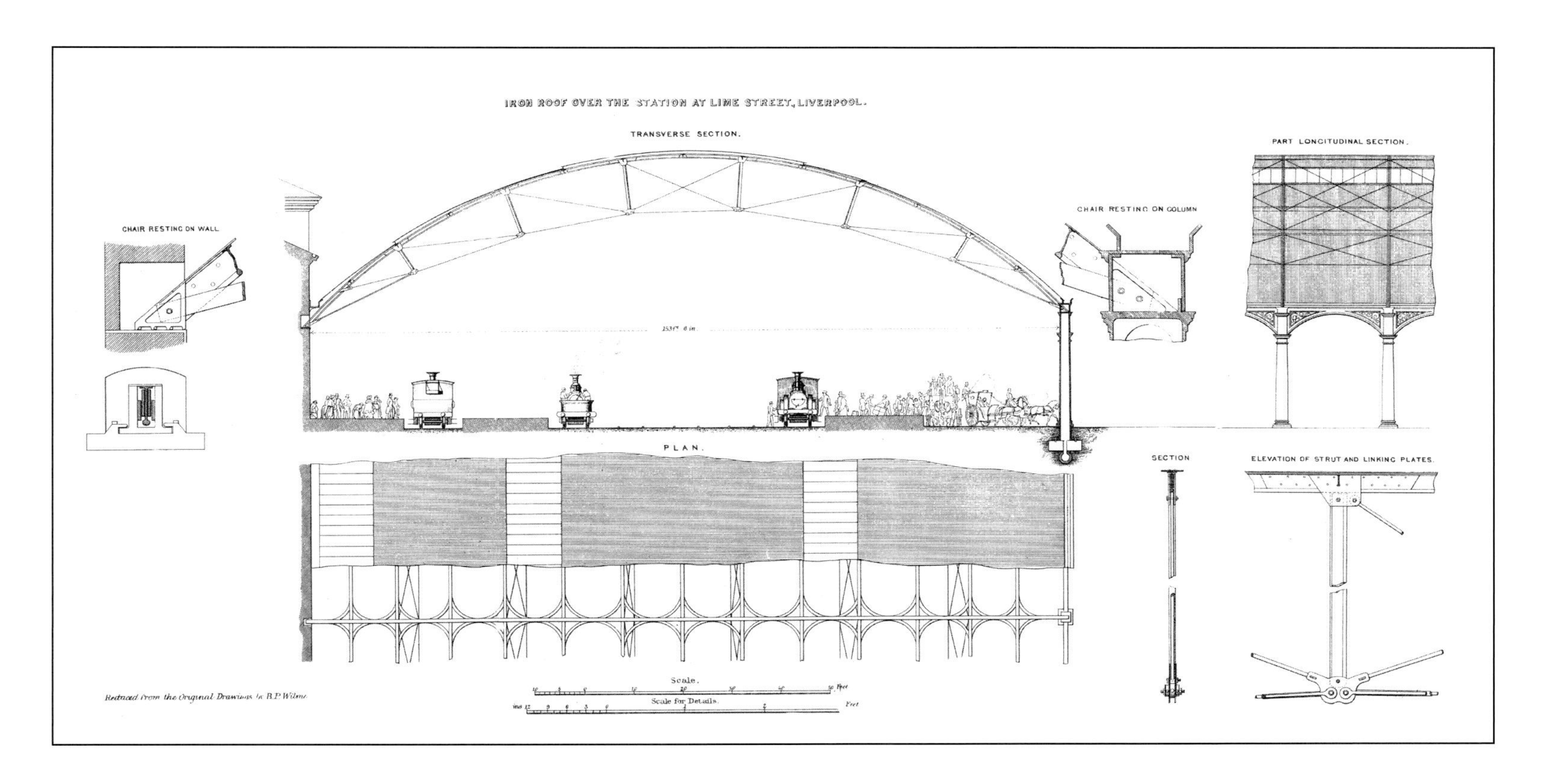

Birmingham New Street Station, previously known as 'Grand Central Station', shortly before the outbreak of the First World War. The station was completely rebuilt in the 1960s.

still problems, particularly with galvanizing, and techniques for purifying the raw material remained ongoing for decades. But while a good deal was now known about the material, it should be emphasized that it took a long time for interest in corrugated iron to build momentum. An 1839 advertisement by Richard Walker, still the sole manufacturer of corrugated iron, listed no more than twenty-five completed buildings, and this was a full ten years after the material had been invented. Nevertheless, Walker was making progress. The buildings that he had roofed in corrugated iron were generally large, prominent structures; many of them were for public use.

Perhaps the key turning point in the early development of corrugated iron was the expiration of Walker's patent. From April 1843 the field was open to all comers. By the end of the decade, driven by increased competition and expertise, the material was primed for take off. Corrugated iron was cheaper than it had ever been, and methods of mass production were well known. All that was required was a mass market.

CHATHAM NAVAL DOCKYARD

Chatham, Kent, England

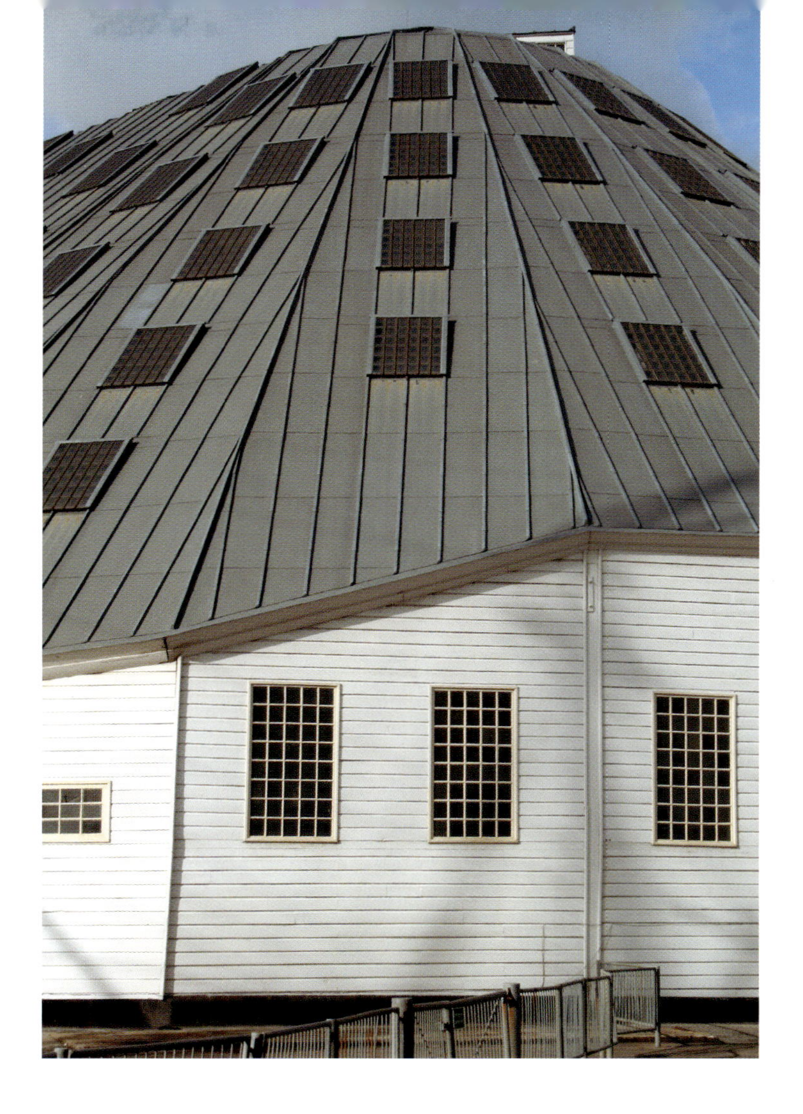

During the 1840s the slips where the British Navy's ships were built evolved from their timber origins into iron structures. The process involved experimentation, mimicry of precedents and finally a confident use of the new material's structural and aesthetic qualities. Nowhere is this progression more apparent than in the Naval Dockyards at Chatham in Kent, five of whose slips built over a fifteen-year period demonstrate three different approaches to design.

Chatham's Number 3 Covered Slipway, completed in 1838, is one of the last survivors from the pioneering age of timber slip roofs. The 100 feet span was based on the innovative ideas of Sir Robert Seppings: a complex arrangement of main frames carries a heavy timber superstructure supporting the roof, whose original surface was made of timber boards and zinc sheeting. It was at the leading edge of structural engineering. The trouble was that it was built of timber, a rot-prone fire hazard.

So in 1844, when work began on the new generation of slips – Numbers 4, 5 and 6 – the decision was taken to build them entirely of iron. While George Baker & Son, a contractor with recent experience of erecting iron slips roofs at Portsmouth Dock, set to work on excavation and the construction of the iron frame, drawings and specifications were produced for the new corrugated iron clad roofs – it is not certain who designed the roofs.

In most key respects Baker's slips at Chatham replicated the system of large cast-iron arches used at Portsmouth. However, the design of the principal roof frames there was not as innovative as the structure they supported. In response to the minimal weight of the corrugated roof covering, Baker & Son spaced their purlins 30 feet apart. This approach was in contrast to the work of Fox Henderson, recognized experts in the design of iron roofs, and Baker's principal competitors.

In their work at the Pembroke dockyard, Fox Henderson had effectively translated the main wooden frames devised by Seppings into well-engineered wrought-iron structures. These frames across the slips were, however, spaced unadventurously, at about the same pitch – or even less – than the wooden ones based on the Seppings model. At Chatham, Baker & Son took things a step forward. From that point on, wide spacing of principal frames became normal for big sheds.

RIGHT TOP Number 3 Covered Slipway, based on the innovative ideas of Sir Robert Sepping, is one of the world's few surviving large-span timber-framed slips.

RIGHT BOTTOM The complex supporting framework of the Number 3 Slipway. By the 1830s timber had been pushed to its limit.

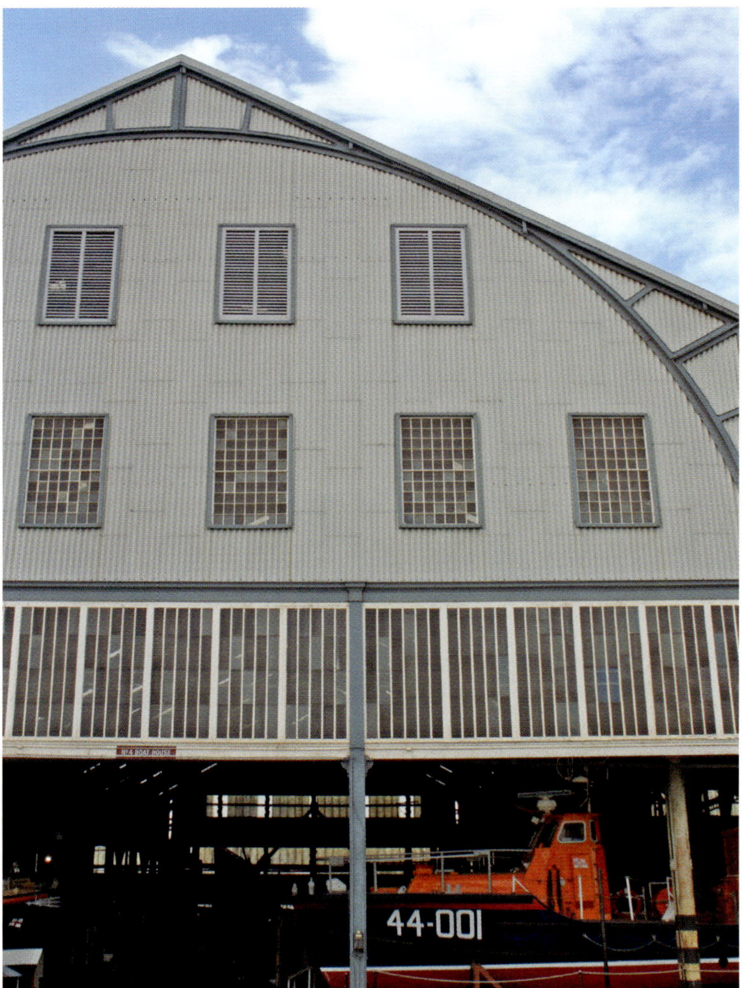

ABOVE AND LEFT The use of wrought-iron framing and lightweight corrugated iron roof panels transformed the appearance of slips 4, 5 and 6. This new generation had wide-spaced principal frames and unencumbered roof structures.

These new slip roofs, although still standing and in good order more than 160 years after they were built, have not survived without some structural problems. Early in 1850 a potentially disastrous collapse of part of the foundations placed enormous strain on the roofs and led to some damage. An enquiry ensued and it was found that Baker & Son was at least partially to blame, although the pioneering nature of the project led the Admiralty to take a lenient view by accepting a share of responsibility.

These shortcomings in the design and construction of the slip roofs were included in the considerations of the Chatham officers in 1851 for a new slip, Number 7, designed by Godfrey Greene, which was another new type of structure. It was among the first to use an iron structure without looking back to earlier precedents. It also served a different purpose. No longer were the supports of the roof required simply to carry the weathering surface of the corrugated iron. Instead they held up tracks for overhead cranes in what was now, essentially, a shipbuilding factory as components became heavier and iron itself entered the construction of Naval vessels.

Number 7, completed between 1853 and 1855, provided an 82 feet wide nave, about 65 feet high with 25 metre-wide chancels for storage and workshops. Iron framing supported a roof of galvanized corrugated iron sheeting with glazed panels and a full-length glazed lantern.

The building had a long and very successful career, although the increasing dimensions of battleships meant that by the 1890s the whole hull could no longer be built completely under cover. By then the other slips were approaching the end of their days. In 1901 plans were announced to convert Number 3 into a boat store. The conversion of Numbers 4, 5 and 6 followed shortly afterwards.

The last ship to be built at Chatham, the submarine *Okanagan* for the Canadian government, was built in Number 7 and launched in 1966. Between 1966 and 1981 Chatham briefly took on a new specialist role, in which Number 7 had a small part, as a docking, refuelling and refitting base for nuclear powered vessels. In 1981 it was announced that the yard would close and since 1984 it has been run as a heritage museum.

ABOVE From timber to iron. The Chatham slips, built during the 1840s and 1850s, reveal an increasingly confident and sophisticated use of iron in the construction of large sheds.

BELOW Number 7 Slipway, completed in 1855. An iron frame supports a roof of galvanized corrugated iron and glass panels.

PADDINGTON STATION

Paddington, London, England

LEFT Isambard Kingdom Brunel used the inherent stiffness of corrugated iron to create a 180,000-square-foot roof without the need for purlins or tie rods.

OPPOSITE Drawings of the 'support rib and 'stressed skin' roof design, which helped to eliminate a layer of structure. The illustration also includes a detail of Matthew Digby Wyatt's arabesque design, the product of Brunel's request for 'detail of ornamentation'.

The role of Isambard Kingdom Brunel in the evolution of corrugated iron has rarely been acknowledged, perhaps because his other achievements were so colossal.

In the 1830s he became aware of the potential of the new material. By the 1840s he was at the hub of a creative milieu experimenting with techniques for its development and production. And in the 1850s he brought fresh ideas to its use of the material on a grand scale in his characteristically creative design for Paddington Station. The cathedral-like volume stands testament to Brunel's powers of invention. It is a landmark of structural insight, and it would not have been possible without corrugated iron.

The rail industry was still in its infancy in the 1830s, and Brunel's exposure to it limited to a single trip on the Liverpool & Manchester Railway in December 1831. Nevertheless, at the beginning of 1833 the directors of the new Bristol Railway appointed him to determine the route of its proposed new line to London.

That March Brunel began work. Ten weeks later the preliminary survey was complete and this and Brunel's proposals for the line, stations and rolling stock was accepted. The London & Bristol Railway became the Great Western Railway (GWR) and on 27 August appointed Brunel its first company engineer. Two and a half years of preparations followed and then came the construction of the line, with its innovative tunnels, embankments, bridges, cuttings and stations (even the engines were a radical new design). Eventually, in June 1841, the GWR's main line was complete from London to Bristol at a final cost of just over £6 million.

When Brunel turned his attention to the construction of the line's permanent London terminus at Paddington, he was still learning about the advantages of galvanized corrugated iron, but he quickly caught up with advances in the field. In 1844 he set up the Galvanised Iron Company with Charles Tupper and George Carr to explore opportunities arising from the collapse of Craufurd's galvanizing patent the previous year. So it comes as no great surprise that he chose to use galvanized corrugated iron for the new Paddington roofs.

Brunel himself took charge of the choice of the roof cladding, having discussed and specified his choice of material with a representative of the supplier – Charles Tupper – in a meeting at his offices in March 1851. But Brunel, the great experimentalist, was to use it in a completely new way.

Corrugated iron's great advantage was its lightness, allowing builders of very large span buildings to use correspondingly light supporting structures. The roofs designed by Fox Henderson at Pembroke Dock and George Baker & Son at Chatham had incorporated the sheet iron as a light weatherproof covering over an iron frame.

Brunel saw that the inherent stiffness of the new corrugated iron sheets could contribute to the stability of the roof and eliminate a whole layer of structure. Paddington is remarkable for having no purlins (longitudinal beams to support rafters) or, indeed, tie rods, helping to create the soaring and uncluttered atmosphere inside the station.

He designed a building with 180,000 square feet (55,000 square metres) of roof covering, applying the galvanized corrugated iron sheets to the lower part of each roof (covering about two thirds of the surface) in a manner that would stiffen the whole structure and would make this roof a very early example of a 'stressed skin' design.

Plans for Paddington began in earnest in 1850, as traffic on the lines began to increase. It was only in December of that year that the Board

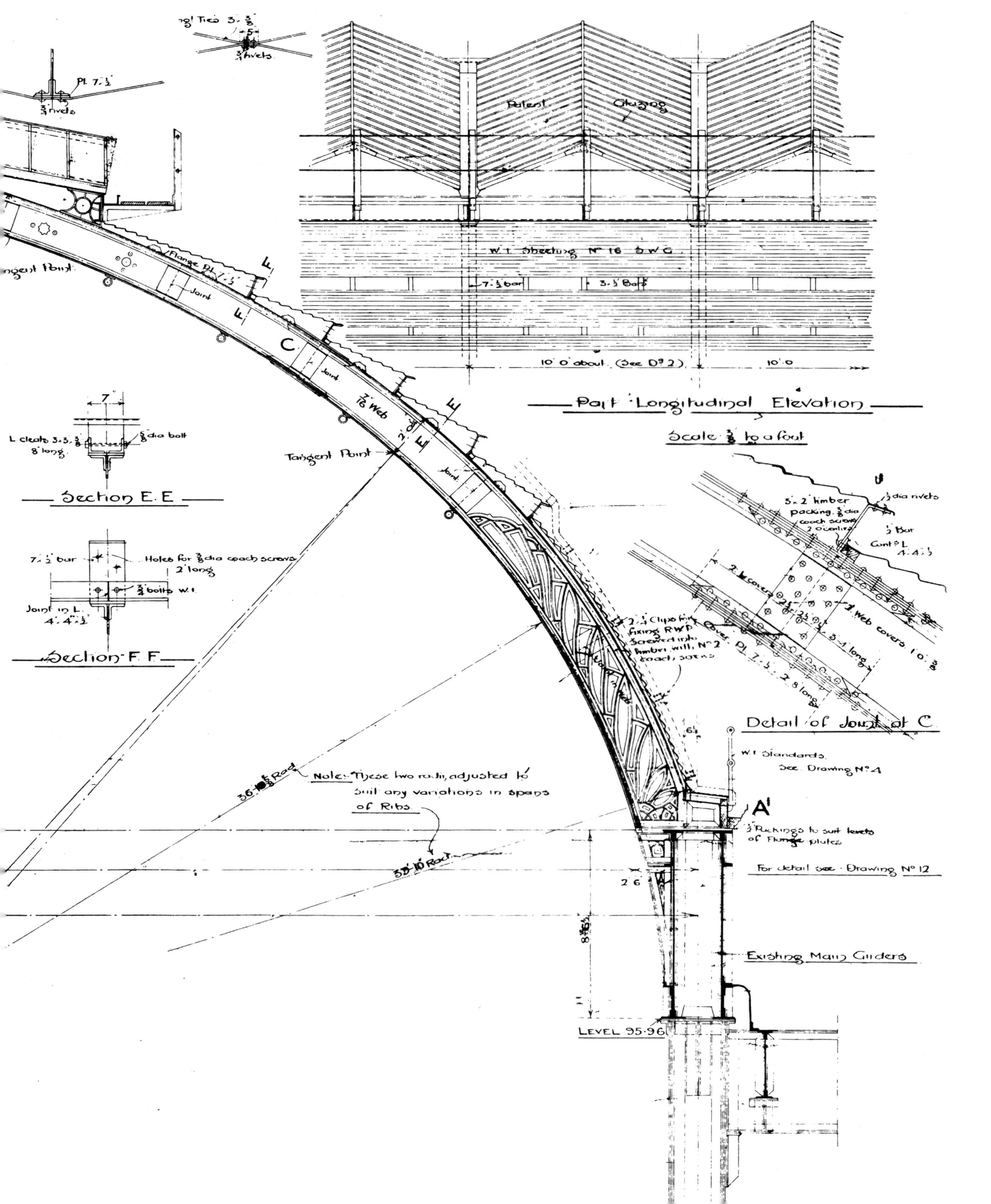
Patent Glazing
W.I. Sheeting No 16 S.W.G.
Part Longitudinal Elevation
Scale ⅜ to a foot
Section E.E
Section F.F
Tangent Point
Joint
Detail of Joint at C
W.I. Standards
See Drawing No 4
Note: These two radii, adjusted to suit any variations in spans of Ribs
For detail see Drawing No 12
Existing Main Girders
LEVEL 95·96

LEFT Galvanized corrugated metal sheets cover approximately two-thirds of Paddington's roof area.

BELOW A little-used corrugated metal profile, of the type used at Paddington, with the corrugations running horizontally.

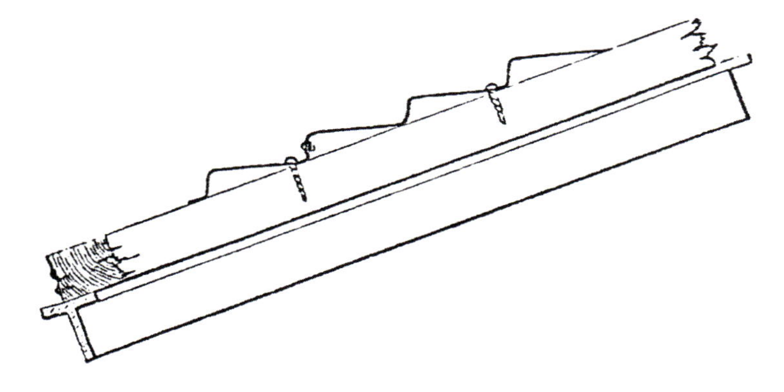

formally gave Brunel the go-ahead. Sketchbooks held at Bristol University Library reveal that Brunel had been sketching his ideas for Paddington as early as 1836. His designs had begun as simple open-sided sheds covering each platform; by early 1851, when work began, those ideas had matured into something substantially different. But one thing that had never changed was Brunel's ambition to build an extremely large and ambitious station as the centrepiece of his project. He also had a dream of giving it a prodigious roof of iron and glass. The final barrel-vaulted design has a central span of 102 feet 6 inches (31.5 metres), the northern span is 69 feet 6 inches (21.4 metres), the southern 68 feet (20.9 metres) and the whole is 700 feet (213 metres) long.

Brunel chose T. A. Bertram – his resident engineer for the eastern half of the new line – to assist him in designing the new station, and wrote to the architect Matthew Digby Wyatt – they had worked together on the early designs for the Crystal Palace – asking for assistance in decorating it:

> I am going to design, in a great hurry, and I believe to build, a Station after my own fancy... it is a branch of architecture of which I am fond... but for detail of ornamentation neither have time nor the knowledge... I want to show the public that colours ought to be used.

By the end of January 1851 the Board had accepted plans for the first of the buildings at Paddington, and Fox Henderson had been signed up as the engineering contractor. There were many delays during the construction process. One of the reasons was that Fox Henderson had severely overstretched itself; the company was also working on the Crystal Palace and the enormous roof of New Street Station, Birmingham. But finally, on 16 January 1854, the first locomotive was run out of the (unfinished) station.

Brunel's great roof remains today, for the most part, as it was built. The northern axis of the station originally opened into a lower aisle that was later (between 1909 and 1914) enclosed by a ridge and furrow roof and is now known as Paddington's fourth span.

The roof has been overhauled numerous times during its life (the combination of steam and sulphurous smoke creating a powerful oxidizing atmosphere under the roof), culminating in the complete reinstatement of the corrugated panels and the replacement of the glass with polycarbonate in 1988–92.

Today it is the only station designed by Brunel still in use.

Samuel Hemming advertisement, *c.*1853. From the late-1840s gold rushes, international trade and colonial expansion helped to create major global markets for corrugated iron, which in turn gave manufacturers the confidence to expand their product ranges and invest in promotion.

2 PORTABLE BUILDINGS

If the two decades following the invention of corrugated iron were about experiment, the second half of the nineteenth century was about exploitation. It was during this period that corrugated iron was harnessed for use in countless diverse applications and emerged as the primary material of a world-wide industrial vernacular, a position that it retains today.

There were obstacles and opposition along the way. The austere industrial appearance of galvanized corrugated iron, and its characteristics of portability, impermanence and ease of construction certainly did not meet with universal approval. From the 1870s it also lost the cachet of novelty; it was no longer new. But in many key respects these were corrugated iron's halcyon years, a time of innovation, growth and acceptance.

Contemporary press cuttings reveal that throughout the 1850s buildings clad in the silvery metal attracted the attention of a broad cross-section of society, not just trades working closely with the material. The interest extended from Britain to mainland Europe, the United States and beyond.

Manufacturers of prefabricated iron buildings had begun to invite the public to view their finest productions prior to export during the 1840s. As well as being a means of promotion, this was also an opportunity to check for defects and to number components. The practice continued into the 1850s, and drew increasingly large crowds. In 1854 an ornate Customs House for the port of Paita in Peru attracted 25,000 people in ten days to its manufacturer's Manchester works (see pages 58–60). The first of six churches designed by Samuel Hemming for Melbourne drew similar interest in Bristol (see page 81). As well as its novel appearance – galvanized metal was still a shock in a world built predominantly of timber and stone – the speed at which it was possible to erect these large structures was unprecedented.

By the early 1850s manufacturers also knew more about corrugated iron. The properties of the material were now well understood and there were few limits to its production on a massive scale. Crucially, the second half of the nineteenth century also saw the evolution of markets that allowed manufacturers to grow, giving them the courage to explore ideas and develop new ways of using corrugated iron.

The first of these new markets was created by gold. From 1849, rushes in California and Australia, followed by hundreds more around the world, generated a substantial demand for prefabricated corrugated iron buildings, as well as individual sheeting. The quest for rare minerals continued for the rest of the nineteenth century, generating enormous popular appeal – it was a romantic adventure open to all. And corrugated iron was intimately tied to those adventures. Sheets, whether new or recycled, occupied a prominent position in many prospectors' temporary settlements.

In Britain the profile and popular perception of the material was boosted by its prominence at London's Great Exhibition of 1851, and its subsequent patronage by Prince Albert. Following the event Queen Victoria's consort ordered a corrugated iron ballroom for the Balmoral Estate in Aberdeenshire. Remarkably, it still stands.

Another legacy of the Great Exhibition was the Kensington Gore Museum, just down the road from the exhibition site. The utilitarian composition of three large vaulted structures was the first major public building sheathed in corrugated iron in Britain. It was a product of the times. No architect was involved in its design. It was a creation of industry for the benefit of science. It was also plagued by operational problems, which only served to fuel the fires of its numerous detractors. The museum barely lasted a decade on its original site before being relocated to the less discriminating East End of London. It was replaced with a conspicuously permanent classical edifice (now the Science Museum).

The negative response to the Kensington Gore Museum reveals much about contemporary tastes, as well as the views of some within the architectural profession, who regarded corrugated iron as a material for which the accepted languages of architecture had no vocabulary and as a potential threat to their livelihood. Corrugated iron remained a popular target for attack throughout the second half of the nineteenth century. A letter to *The Builder* in 1858, caught the public mood:

> Cannot something be done to improve architecturally the various iron structures we see erecting [sic] by the many eminent firms in the iron trade?... Unless something is done, iron will be superseded by a material... less obnoxious to public prejudice.

SAN FRANCISCO, FROM THE SOUTH-WEST.

A view of San Francisco in 1850. Corrugated iron buildings would have been among the 'bawdy, bustling, bedlam' of tents and timber dwellings.

The Church of England also remained immune to its charms (see Chapter 3), as did William Morris, a founder of the Arts and Crafts Movement, who in an 1890 pamphlet issued by the Society for the Protection for Ancient Buildings railed against the material, 'now spreading like a pestilence over the country'. But corrugated iron's detractors could do little to stop its inexorable spread.

As the nineteenth century progressed the benefits of corrugated iron became clear to entrepreneurs and innovators engaged in all manner of activities related to housing, trade, industry, agriculture and commerce all over the world. The material was used in railway stations, lighthouses, coaling stations, gas production plants, customs houses and other dock facilities, warehouses, hotels, factories, stores and in a myriad other applications. Some of the early manufacturers even envisaged whole towns of corrugated iron.

It was not just in the pursuit of riches that corrugated iron played an indispensable role. The material was also used for buildings in earthquake and hurricane-prone regions – securely attaching corrugated iron sheeting to framed structures made them less likely to collapse in extreme conditions. Prefabricated hospitals, to isolate people with communicable diseases, or to treat victims of war or natural disaster, were another lifesaving innovation. They could be built rapidly, and were easy to keep clean.

In all of these and other applications, corrugated iron proved itself to be a versatile, cheap and easy-to-use alternative to other materials on the market. And as demand increased, manufacturers proved themselves equally versatile in marketing their products.

Gold rushes

It was the Californian gold rush that gave wings to the nascent corrugated iron industry. Manufacturers responded inventively by redefining their products and tooling up to meet demand. The first Californian gold was found in a tributary of the Sacramento River in January 1848. Within months word had spread around the world. By the end of the next year over 80,000 gold seekers had arrived in the isolated territory. San Francisco in 1849 was described in a contemporary account as, 'a bawdy, bustling, bedlam of mudholes and shanties'. Ready-made wooden shelters arrived from Germany, France, China, Hong Kong, Tasmania and New Zealand. Iron buildings and corrugated iron sheeting also began to appear in California.

The earliest corrugated iron houses to arrive were shipped from New York, via the Isthmus of Panama and Cape Horn. A metal roofer, Peter Naylor, was among the most successful manufacturers. In 1849 he shipped between 500 and 600 corrugated iron houses – measuring 20 feet by 15 feet – to the goldfields from his works in Stone Street, New York. In an advertisement of February 1849, Naylor described both his system of construction and the means of transport to California:

> The Galvanized Iron House constructed by me for California having met with so much approval, I am thus induced to call the attention of those going to California to an examination of them. The iron is grooved [sic] in such a manner that all parts of the house, roof and sides slide together, and a house... can be put up in less than a day. They are far cheaper than wood, are fireproof and much more comfortable than tents. A house... can be shipped in two boxes, 12 feet long and 2 feet wide and 8 inches deep, the freight on which would be about $14 to San Francisco.

Aside from Naylor and some Belgian suppliers, the majority of manufacturers known to have shipped corrugated iron cottages and warehouses to the Californian gold fields were British. Morewood and Rogers, and John Walker, the son of Richard, who set up his own manufacturing operation in 1849, were among the first to capitalize on gold fever. That August *The Builder* reported that John Walker had sent a large

ABOVE In 1849 Liverpool engineer John Grantham exported corrugated iron warehouses to San Francisco. He also supplied the American army with barracks.

LEFT An 'Iron Store-House' designed by John Walker for export to San Francisco.

BELOW Two-storey structures with commercial premises above residential accommodation, and houses with up to 12 bedrooms were among the range of corrugated iron buildings designed by Edward T. Bellhouse.

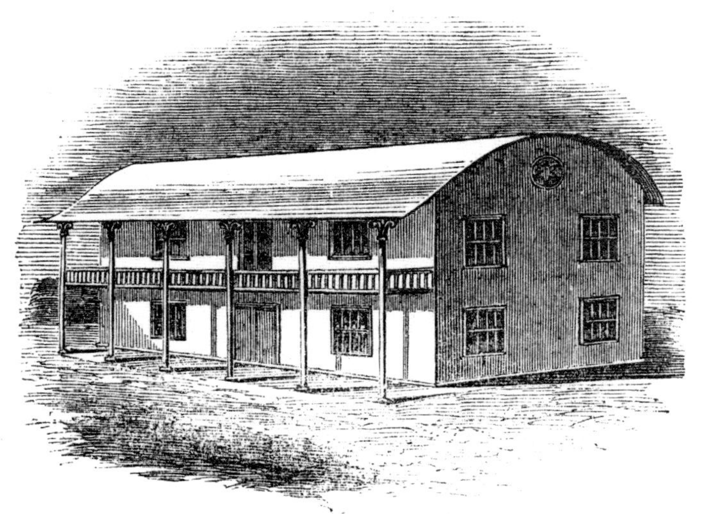

IRON STORE-HOUSE SENT TO SAN FRANCISCO.

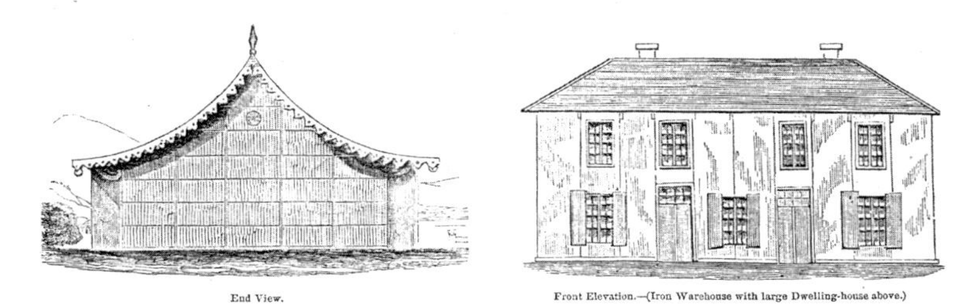

End View.

Front Elevation.—(Iron Warehouse with large Dwelling-house above.)

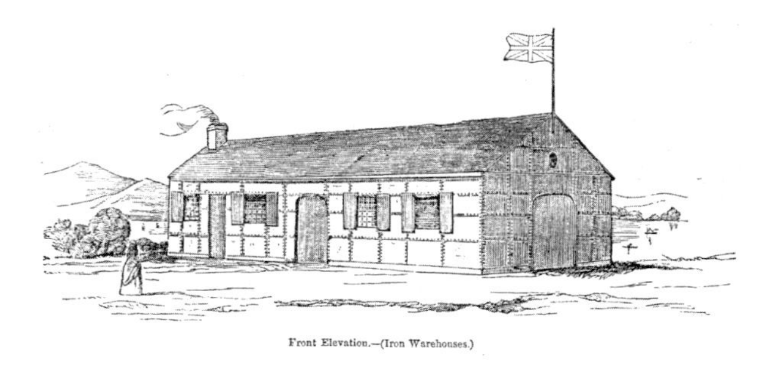

Front Elevation.—(Iron Warehouses.)

two-storey warehouse, constructed of eight-feet long corrugated iron sheets and costing £600, to San Francisco. Another eight buildings destined for the gold fields, each consisting of three living rooms and a store, were also under construction at his Old Street Road works in London. In Liverpool, Vernon and Son's Brunswick Dock shipyard was another site of frenzied activity. It was there that John Grantham, a local civil engineer, built an iron warehouse for California in just 23 days. The speed of construction and scale (110 feet long by 30 feet wide by 20 feet high) impressed all who saw it. The *Illustrated London News* remarked: 'The whole of the iron was galvanised, it was nearly white and had a singular appearance.'

Manchester engineer and iron founder Edward T. Bellhouse, an early innovator in the design of prefabricated iron buildings, was also swift to respond to the Californian gold rush. From 1849 he sent several hundred houses to the American West coast, some were fitted with carpets and wallpaper. His range included cottages or cabins that cost £40–£80 each, larger houses of up to twelve rooms that cost around £1,000 and warehouses. All were made using a system of 6 feet by 2 feet corrugated sheets on a T-section iron frame.

The combined output of British corrugated iron manufacturers was prodigious and attracted the attention of a Viennese publication, which reported: 'Iron is cheap in England so people thought of using it to make houses. . . they have already made hundreds of such and sent them to California, some completely furnished.'

EAST INDIA VILLA.

HEMMING'S PATENT IMPROVED PORTABLE HOUSES,

SOLE MANUFACTORY, CLIFT HOUSE, BEDMINSTER, BRISTOL.

THE Houses consist of an Iron or Timber framing, if of Timber 4½ by 3 inches, or 3 inch square scantling, and sills of Oak, or Memel; in storied Houses the scantlings are increased in proportion.

The walls and roof are of galvanized corrugated Iron, the ridge capping of the same material.

The walls are lined in the inside with ½ inch boarding, covered with canvass, ready for papering, leaving a space of 4½ or 3 inches throughout the entire building between the Iron and the Wood-work, afterwards to be fitted up with any non-conductor, by which means a complete ventilation is effected, and the temperature in Summer much lessened, and increased in Winter; this place is adapted for 4½ or 3 inch brick-work.

The ceilings are canvass with felt, ready for papering. The space between the Felt and the Galvanized roof to be filled up with any non-conductor.

The doors are pannelled, and furnished with good locks and hinges.

The sashes are glazed with glass 21 ounces to the foot, with shutters and fastenings complete. All necessary skirting boards, architraves, and other mouldings are provided. The whole of the Wood-work usually painted, has four coats in oil.

The erections are entirely put together with iron screws, and completely erected at the works previous to shipment, and may be re-erected by any inexperienced person in a few hours, every part having been carefully fitted, numbered, and lettered; descriptive plans being given with every building.

Every description of DWELLING HOUSES, STORES, SHOPS, WAREHOUSES, CHURCHES. CHAPELS, SCHOOL-ROOMS, of any dimensions, to any given Plans, may be constructed in a few days ready for Shipment, delivered free alongside in Bristol, or at the Bristol Railway Stations.

AN EMIGRANT POPULATION WITHOUT LODGINGS; THE EVIL AND ITS REMEDY.

By a recent Visitor, Reprinted from the "Lady's Newspaper."

EMIGRATION may be of two sorts;—the one a reckless and ignorant desire of change, full of present discomfort, and ending most commonly in expectations disappointed and feelings disgusted;—the other, the calm determination of sober reflection, that weighs its peculiar difficulties, and wisely sets itself to the task of overcoming them. The one class of Emigrants are like the Tourists who would go forth on a long journey unprovided with umbrella or great coat, because the sun shone when they set out; the others are the careful travellers, who have these appliances carefully folded up and stowed away for use when wanted.

It should never be forgotten by the Emigrant that he is going from a country of cheap labour, to a land where labour is dear; and this, which is one of the main inducements to Emigration, should lead him to provide at home those indispensable necessaries which he must otherwise procure, at whatever cost, on his arrival. Among the first of these necessaries we must surely reckon a shelter—a home; and though in a climate like Australia, we may take liberties which elsewhere would be fatal, yet it is wise to recollect, that even to the healthy and vigorous man, a fit of sickness and long season of debility is too often the penalty he pays for determining to "knock along somehow," to say nothing of the exorbitant price which the meanest pig-stye commands as rent, and which soon exhausts a deeper purse than would have supplied a comfortable home for the stranger, keeping him both in health and spirits.

The last time we met MRS. CHISHOLM, she very strongly expressed her conviction, that persons taking out moderate sized houses would have the comfort of them for months after landing, and finally sell them at a very handsome profit, after paying freight, and all expenses of carriage. The personal interest of this lady in the Colony, led her to pay two visits of inspection to Clift House, and we may rest assured that hers was not a mere lady's glance of approbation at every novelty that came under her notice, but a business-like examination and inspection of detail; the result of which was an order for several houses.

The manufacture of houses for the Australian Colonies is therefore become an important branch of industry, and the successful Inventor has found many to *imitate*, who were unable to *originate*. In our first visit to the Factory, at Clift House, we were very forcibly struck with the *genius* of the proprietor: we felt that there was a master-mind at work, in the careful adaptation of means to the end required Nothing superfluous—Nothing wanting—Nothing left unthought of, to be remembered when it was too late to remedy. As it was a father's thoughtfulness for his son that suggested the invention, and originated the Establishment, nothing was left unprovided for his comfort; and it has seemed as if the same spirit had actuated and prompted the multitudinous adaptations of the principle to every conceivable variety and style of building; and it is this which not only gives a charm to a visit to Clift House, but makes the interest of the Emigrant safe in the Proprietor's hands; for he has only to explain the object he wishes to accomplish, when the same genius which suggested the first fabric, will be exerted to meet his particular want.

We happen to know of one very sagacious party at a distance who fancied that it was a simple thing to put together an Iron house; and as he had a quantity of materials by him, he thought nothing further was needed than an inspection of the buildings at Bristol. He forthwith came down, and Mr. Hemming with the urbanity which ever prompts him to give his time to visitors, escorted the stranger about, and with much minuteness explained the various particulars of his erections, all unconscious that instead of a customer he was arming a lynx-eyed rival with weapons to use against himself. The gentleman went home fully satisfied with the information he had gathered, put up his house, and found to his astonishment, that its cost, exclusive of the materials he had himself furnished, was more than Mr. Hemming would have charged for a better erection, including the carriage.

The church ordered by the Bishop of Melbourne, is pretty familiar to most parties from the Engravings, and so satisfactory has it proved in its complete state that a second was erected for the same parties, to give 1,000 sittings together with a Parsonage, more roomy than the comfortable residence prepared to go with the first church. The Hotel, which has been purchased by an enterprizing Capitalist, (and which we have no doubt will prove an excellent Investment,) will make up 50 beds, each provided with means security for the occupants luggage. One is now in course of construction on a much larger scale. Indeed the means of accommodation in the Colony are so disproportioned to the increase of population, and must continue to be so, that a Building costing in Bristol about 1,800 guineas, will at a fair Colonial rental, return £2,400 per Annum. We have seen a row of 2 storied houses and shops which would not disgrace Hastings or Malvern; and where the intended site is a town, the emigrant would do well to consider the cost of ground, and increase rooms upwards instead of sideways.

The first idea that strikes a stranger, and it is one that interested parties, we are sorry to say, are not slow to encourage, is that these buildings must be both insufferably hot and miserably cold, and that however neat and tasty they may look in the Builder's yard, after a little Australian sun-shine, they will let in more light than comes through the windows. Now all this we conceive would most likely be the case with an Iron house put up by a common builder, but Mr. Hemming is not simply a builder; and a residence in tropical climates has taught him both what to provide for, and what to provide against. The shrinkage and warpage is prevented by subjecting all the timber, previous to use, to a seasoning apparatus on the premises; while a clear space of several inches betwix the Iron and Wood work of the buildings, and a wide space between the ceilings of papered inodorous Felt, and the Roof, secures an ample ventilation and equable temperature. We have seen children amuse themselves with ingenious toys of houses, churches, &c. which they put together by the aid of a model card, and just so are these habitations for living beings; but while the one is the amusement of a child, the other is destined to save the lives and add to the comforts of thousands of our countrymen. Nor will they be confined to Australia. We have seen a very beautiful Villa, intended we believe, for the banks of the Rhine, and we can conceive of numberless instances in this country where the advantages of an erection, economical, durable, and that is strictly a Personalty and removable from another party's land, will bring them into requisition.

OPPOSITE A promotional poster by Samuel Hemming describing the benefits of iron lodgings to 'emigrant populations'.

ABOVE A 'Villa Residence' designed by Samuel Hemming for Messrs B. L. Lloyd & Co. of Sydney *c.*1853.

LEFT A page from a Charles D. Young & Co. catalogue *c.*1857. Almost identical structures, credited to E. T. Bellhouse of Manchester, were published closer to the California Gold Rush.

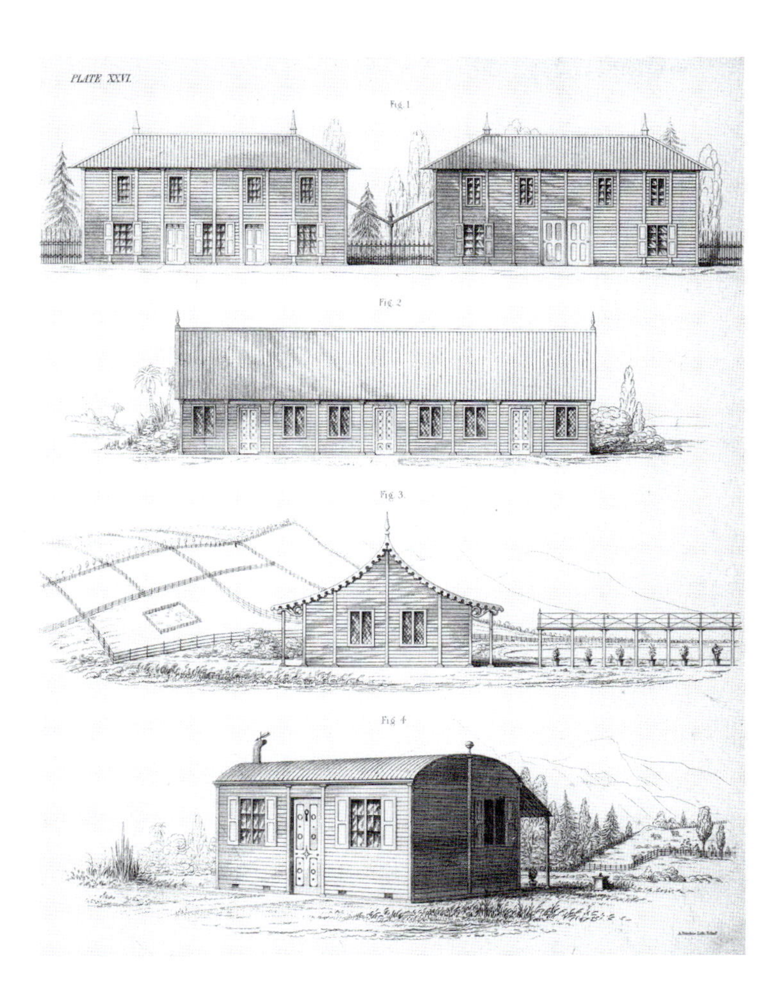

The Californian market dried up almost as quickly as it had started. In October 1849 houses were still selling for ten times their cost price, but by March 1850, entrepreneurs had begun to build sawmills, so providing a local source of building materials. Within months corrugated iron houses shunned in California were being sold at bargain prices in Hawaii.

The momentum behind the manufacture of corrugated iron buildings was dissipating quickly, but towards the end of 1851, gold was found at Lewis Ponds, New South Wales and Clunes in Victoria. British manufacturers of corrugated iron were quick to respond to the new markets. At the time, buoyed by the growing wool trade, the sea-lanes to Britain's colonies in Australia were full of traffic. So there was plenty of unused cargo capacity for the outward journey to Australia. Indeed, many captains actively welcomed cargoes of corrugated iron, a relatively heavy and compact material, as ballast.

One of the first to respond to the Australian rushes was Samuel Hemming, the Bristol manufacturer, who set up his Clift House Works in 1853 and immediately won substantial business (see also Chapter 3). Crucially, he also secured the patronage of Caroline Chisholm, a campaigner for better conditions for immigrants.

ABOVE In the 1850s South Melbourne was a shanty town of tents, improvised shelters and corrugated iron kit buildings. The only gold rush era prefab still standing on its original site is at 399 Coventry Street.

BELOW In 1858 Colonel Richard C. Moody was dispatched to the new colony of British Columbia to survey the territory and help maintain order during the Fraser Canyon gold rush. His corrugated iron house, designed by Samuel Hemming, is pictured here in 1939, in use as a printer's warehouse. It still stands today.

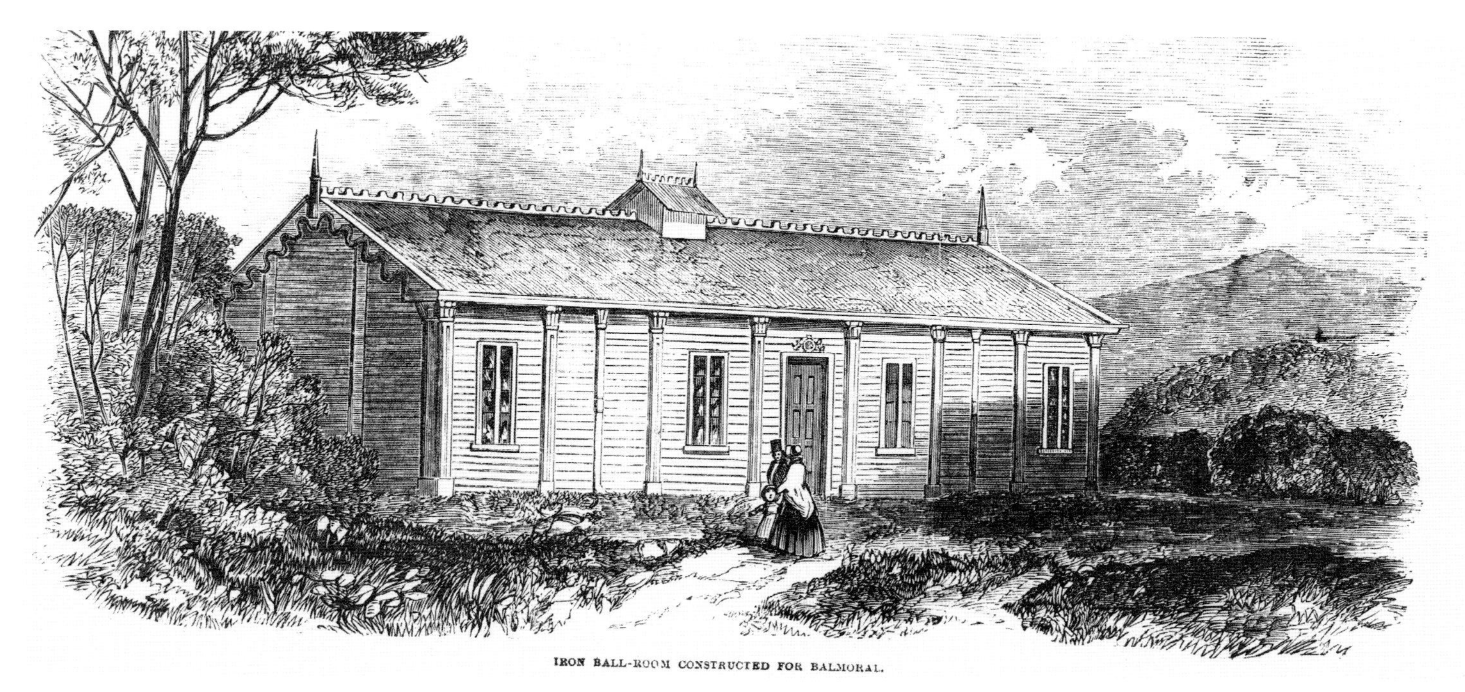

ABOVE The former ballroom at Balmoral Castle, Aberdeenshire, may be the world's oldest surviving corrugated iron building. Prince Albert commissioned it following the 1851 Great Exhibition; it was complete a few weeks later).

For a few years this relationship proved to be of great mutual benefit. The Australian rushes also brought some new faces to the scene. Charles D. Young and Co. was established in 1847 when Charles Denoon bought out his brother William from a business they had set up to produce ironmongery and wire. C. D. Young quickly became influential in the production of cast and corrugated iron buildings. In 1852 it published catalogues of its output, at least one aimed specifically at the Australian market, and others aimed at more general audiences. John Walker was also active in supplying the Australia market. On 2 July 1853 *The Builder* reported that he was, 'in [the] course of constructing thirty-six iron houses for the residences of emigrants sent out by the Government to Australia'. Walker also advertised in the Australian press; his range went from two-bedroom cottages to large mansions costing £5,000.

Of the hundreds of corrugated iron cottages sent to Australia from the 1850s only a handful have survived. Three are owned by the National Trust and are open to the public in Melbourne.

As today, many vessels docked at Port Melbourne, a short distance south of the growing town's centre. When the gold rush began, the land between became an informal community of tents, temporary timber structures and portable corrugated iron houses, one of which still stands on its original site on Coventry Street, South Melbourne. The gabled single-storey house, with an attic, cost £60. It was originally one of five erected on the street by Robert Patterson, a private speculator – the stencilled initials 'RP' are still visible on some internal walls. The five houses – completed in 1854 – were probably designed by Robertson & Lister, a Glaswegian company that exported a number of iron buildings to Australia during the 1850s – the cast-iron window frames bear a strong similarity to those at the Brown Brothers Iron Store in nearby Geelong, a building positively identified as a Robertson & Lister design (see pages 54–57).

The other two buildings owned by the National Trust were relocated from other sites in Melbourne. One is a small cottage designed by E. T. Bellhouse, whose external shell has survived in good order. The other, a larger house thought to have been designed by Morewood and Rogers, may once have housed two families either side of a central corridor. Unusually, some of the internal walls are made of corrugated iron – walls of timber match-boarding were much more common. It was last occupied in 1976.

Rushes for gold and other minerals continued throughout the nineteenth century. There were notable gold strikes in British Columbia (the Fraser Canyon rush began in 1858) and New Zealand (the Central Otago rush, 1861), and in 1870 a major diamond rush broke out in Kimberly, South Africa. Wherever minerals were found, fortune hunters would follow, and for many of them home would be the sheets of corrugated iron that they had brought with them.

The Great Exhibition

For Britain, one of the defining events of the mid-nineteenth century was the Great Exhibition of 1851, an international festival of industry and celebration of Britain's manufacturing pre-eminence. The event was initiated by Prince Albert, a prominent supporter of science and industry, and it was a great popular and financial success.

As mentioned above, the most beneficial outcome of the exhibition for manufacturers of corrugated iron buildings was the endorsement of Prince Albert. Several corrugated iron manufacturers exhibited their wares in Joseph Paxton's colossal Crystal Palace, including Morewood and Rogers, Tupper and Carr and C. D. Young & Co. But it was a model of an emigrant's cottage designed by E. T. Bellhouse and Co.

that caught the prince's attention. Shortly after the exhibition he ordered a similar structure for Balmoral, a royal residence in Aberdeenshire then under construction.

The 60 by 24 feet building was Bellhouse's No.1 Warehouse pattern, which his promotional material described as, 'a spacious structure of corrugated iron plates' with two doors and eight windows, and a ventilator resembling an ornamental chimney. Completion of the ballroom was rapid. Within a few weeks of receiving the order, Bellhouse wrote to Sir Charles Phipps, Keeper of the Privy Purse, advising him that the building would be ready within a week. Two weeks later the building was sent by train from Manchester to Inverness and erected at its new site on the Balmoral Estate.

The ballroom was widely reported and greatly admired. The *Illustrated London News*, in a romantic mood, described the building when seen from a distance as a 'large cottage orné'. It remained in use for its original purpose until 1856, when Balmoral Castle was complete. The ballroom still stands on the estate, where it is now used as a joiner's workshop.

The other major corrugated iron building erected as a direct result of the Great Exhibition was the Kensington Gore Museum – also known as the South Kensington Museum. In terms of durability and positive press coverage its experience was the polar opposite of the Balmoral ballroom.

The Great Exhibition made substantial profits, some of which were used to finance the establishment of a series of permanent museums and galleries in South Kensington. Prince Albert was particularly supportive of proposals to build a museum to celebrate scientific advances catalysed by the Industrial Revolution. In June 1855 the task of designing and building the 'iron house' intended for that purpose was given to C. D. Young and Co. – no architects were involved. Plans of the structure were published in *The Builder* in May the following year. Young's proposal read:

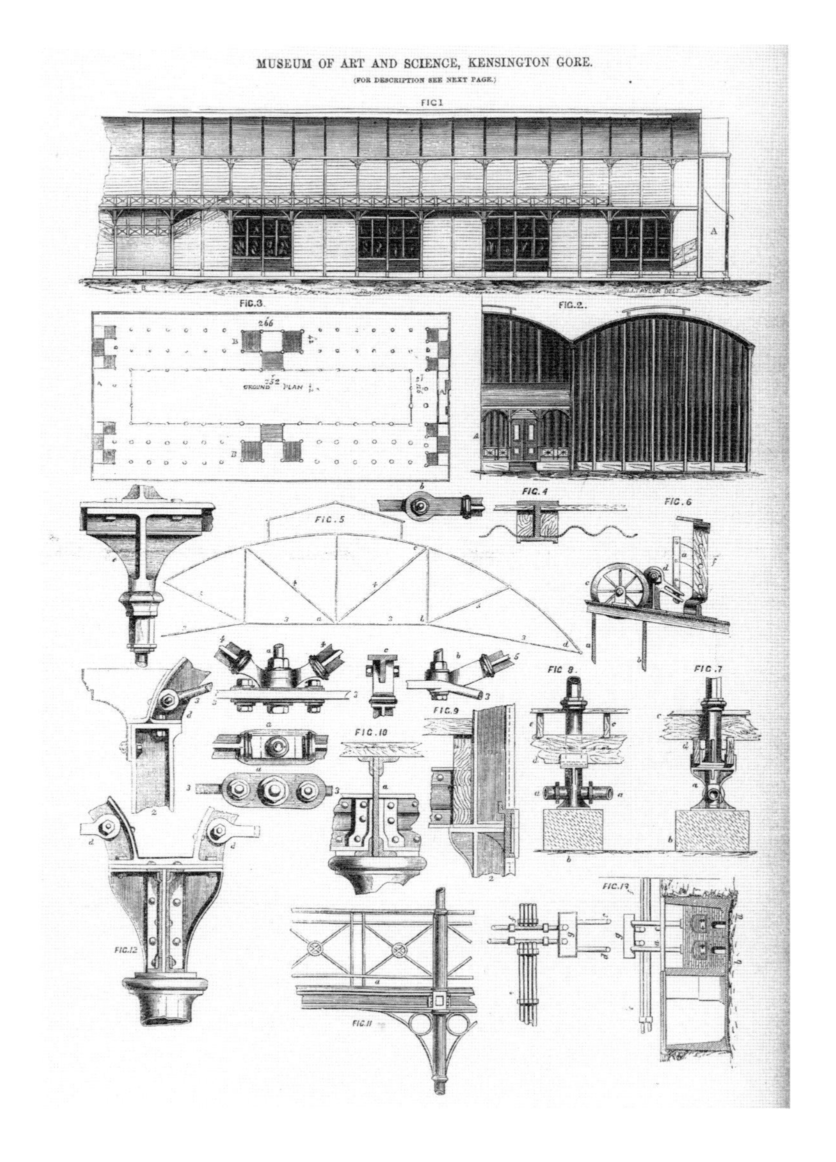

OPPOSITE TOP The brutal rationality of the 'Brompton Boilers' was too much for the tastes of the day. After a decade in South Kensington, the building was relocated to Bethnal Green.

OPPOSITE BOTTOM Detail drawings of the Brompton Boilers.

RIGHT Lighthouse designed by Richard Walker for the Florida Sands (1851), incorporating corrugated iron in the living quarters.

> The building would be in form, 266 feet long, and 126 feet broad and about 30 feet high to the eaves... [It] would cover an... entire space for exhibition of... 1½ acres... The walls of the building would be composed of cast-iron uprights... 7 feet apart... The spaces between the columns would be filled up with corrugated iron sheets... [It] would be covered by three segmented roofs, each 42-feet span, supported on the outside walls ... The trusses would be of malleable iron, 7 feet asunder and covered with corrugated iron sheets... The whole of the ironwork would be covered, within and without, with three coats of oil paint... The cost of the building... would be about £9,800; if with an architectural front of cast-iron from £1,000 to £1,400 additional... The building would be constructed in bays of 14 feet square, or a multiple of that number. By adopting this principle, we obtain greater economy in first erection... or removal and re-erection.

Work began early in 1856 and by mid-April it was sufficiently advanced to attract the attention of the press, and this was when the Kensington Gore Museum's troubles began. Its design was widely disparaged by the editor of *The Builder*:'[The] inartistic iron building was a strange way of demonstrating the value that was placed on art in industry... its ugliness is unmitigated,' he wrote.

Three weeks later, the same publication compared the building to,'huge boilers placed side by side... for it is filled in externally with corrugated iron, and is therefore all the more like a threefold monster boiler'. From that point the museum was known as the 'Brompton Boilers'. Attempts were made to lighten the building's oppressive appearance, firstly by painting it in green and white stripes and, later, by adding the iron portico mentioned in Young's description of the building as an optional extra, but to no avail.

In 1857 an adjacent temporary building, the Refreshment Rooms, stoked up further scorn. On 29 March *Lloyds Newspaper* raged:

> In addition to the hideous corrugated iron structure already erected, and described by the press as a railway shed, or hospital for decayed railway carriages... [the Refreshment Rooms]... being erected nearby... will harmonise with its... neighbour, in as much as it will be hideously ugly... It... will form a contrast to the yellow and green striped erection with which it is so immediately associated.

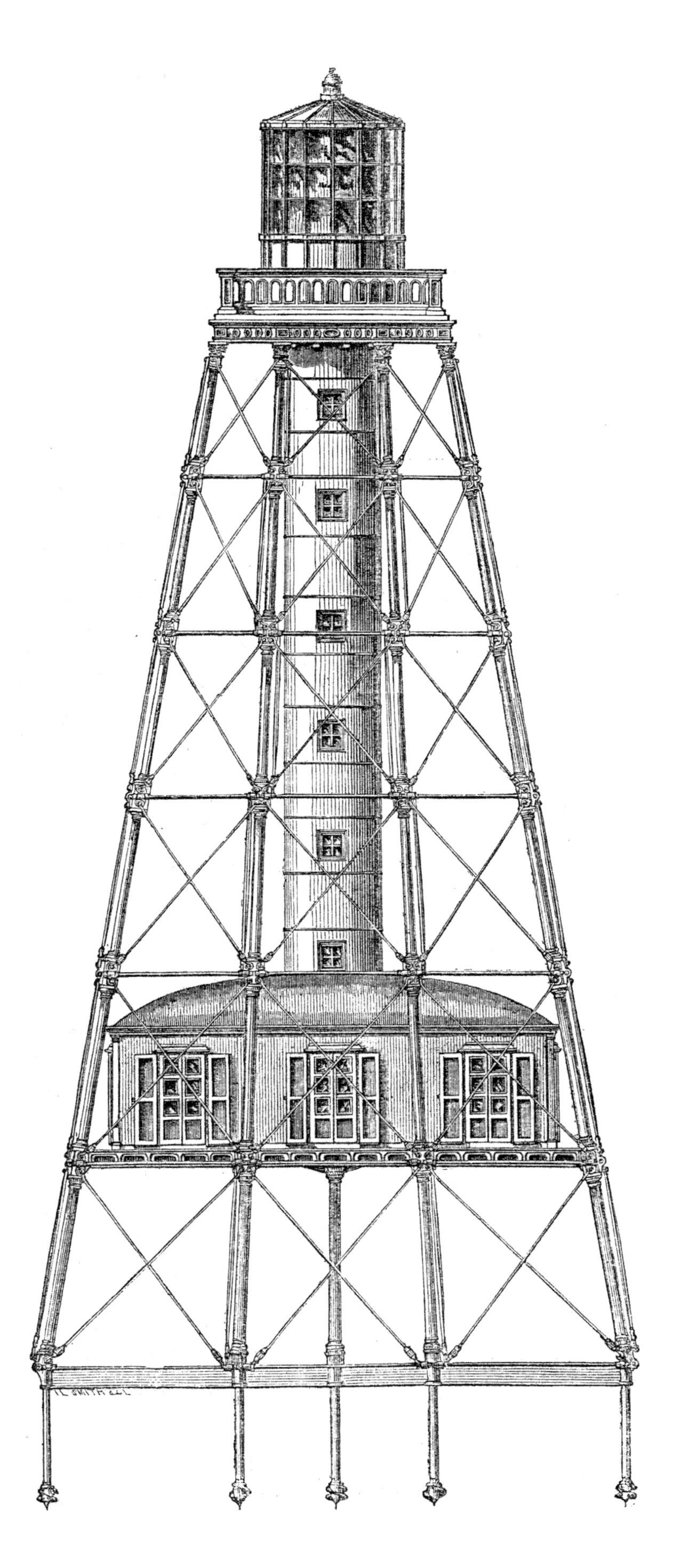

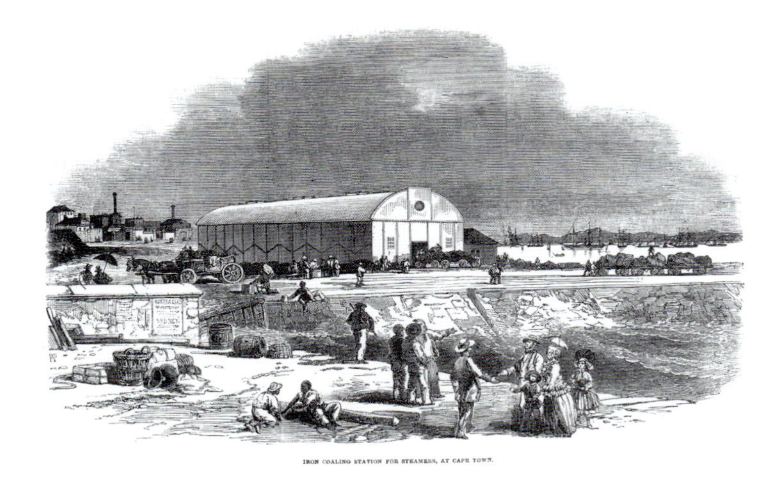

IRON COALING STATION FOR STEAMERS, AT CAPE TOWN.

IRON HOUSE FOR CHAGRES, MADE BY WALKER.

TOP Coaling station in Cape Town, *c.*1854. The large barrel-vaulted shed, designed by Richard Walker, supplied steam ships on voyages to and from Australia.

MIDDLE Premises for the Royal Mail Steam Packet Company at Chagres on Panama's Caribbean coast, designed by John Walker.

BELOW Herd Groyne lighthouse in South Shields, England.

The museum never recovered from its poor start. Prior to its opening on 22 June 1856, it was discovered that the building leaked in twenty-one places; for the nine years that it stayed open, curators had to protect exhibits from the elements. In 1866, with parts of the roof having almost corroded away, the building was dismantled and moved to Bethnal Green, East London. The frame still stands; the corrugated iron enclosing walls have long since been replaced with red brick. Today it is home to the Museum of Childhood.

Transport

Meanwhile corrugated iron was found to be well suited to buildings serving new modes of transport and expanding global trade routes. The growth of railway networks during the nineteenth century is perhaps the most prominent example of the material meeting a need for large numbers of flexible buildings, both small-scale and short-term, and more prestigious structures, like railway termini. But corrugated iron was also used for coaling stations, lighthouses and all manner of buildings associated with the flow of goods and passengers.

Lighthouses were one of the earliest buildings related to transport known to have been built of corrugated iron. The material was perfect for the rapid construction of isolated structures, often in precarious positions. In 1848 Richard Walker designed a lighthouse that incorporated corrugated iron in its living quarters. Another based on similar principles still survives, six miles off the English coast at Frinton-on-Sea. 'Gunfleet' was originally one of three 'Trinity House' designs (the organization responsible for aids to navigation in British waters) built in the Thames estuary during the 1850s. It is supported on seven piles, the central one with a broad-bladed screw twisted into the sand bank upon which it stands. Corrugated iron was used in the construction of lighthouses until at least the end of the nineteenth century. A later example still stands in the Queensland Maritime Museum, Brisbane. It has been relocated from its original site on Bulwer Island at the mouth of the Brisbane River. This modest building is clad in specially made corrugated iron sheets, with tapering flutes to suit the shape of the structure.

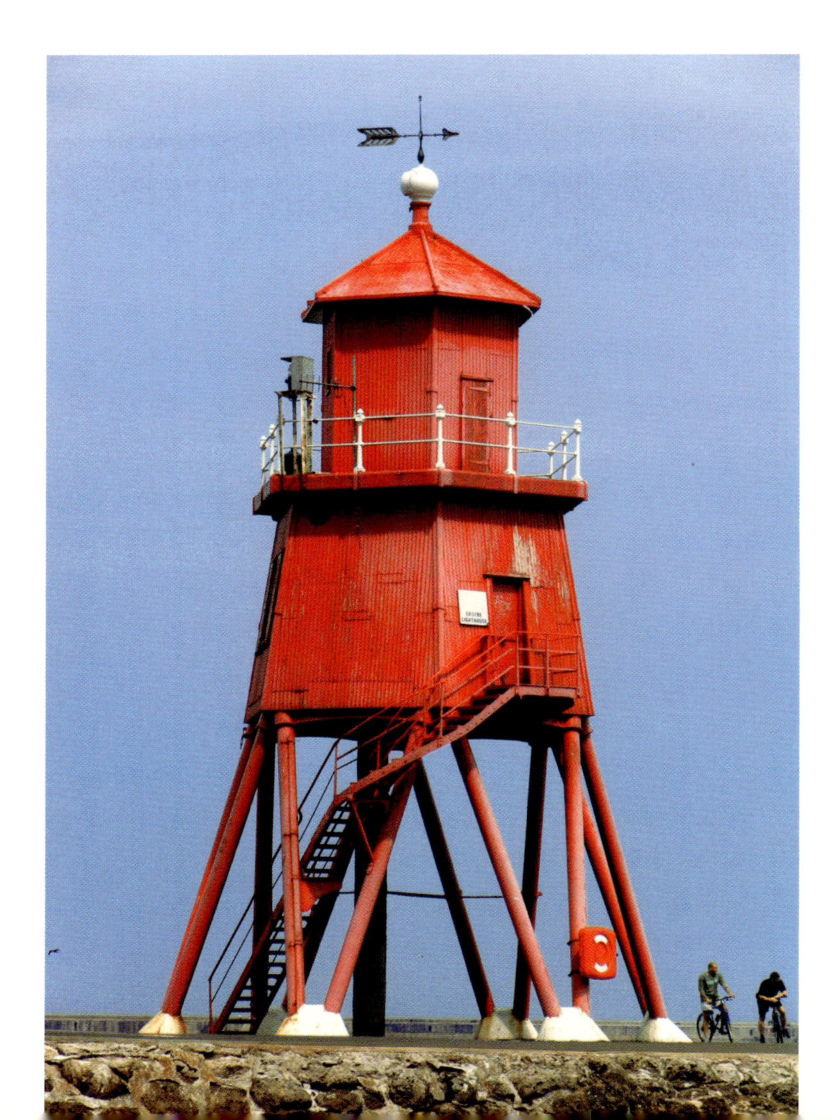

G.W.R. STANDARD LAMP HUTS.

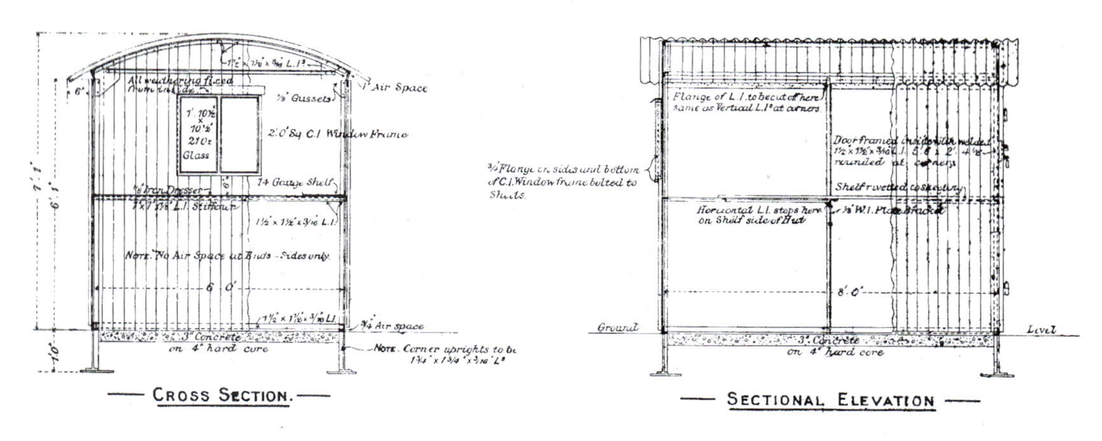

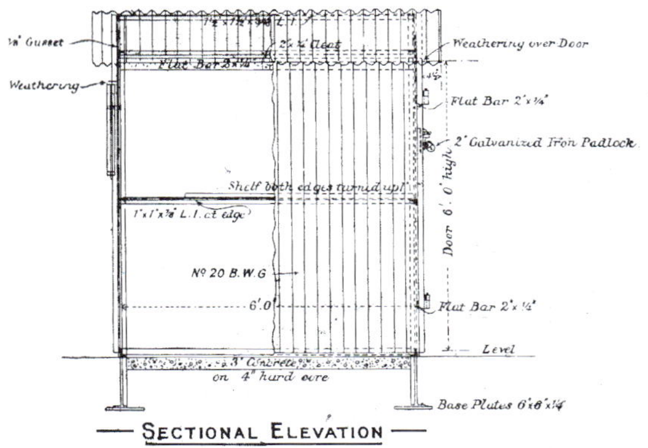

From the 1890s standardized Lamp Huts were built to isolate flammable oil lamps from station buildings on the Great Western Railway. A few have survived, including the hut at Blue Anchor station on the West Somerset Railway, England.

Another, later example is in South Shields, north-east England. The mouth of the River Tyne, as it enters the North Sea at South Shields, is protected by two defensive walls to aid navigation. Inside these protective arms is a small jetty built to prevent the Herd Sands from encroaching into the river mouth, upon which is built, at its sea end, a tall red painted corrugated iron lighthouse raised on cast iron legs. Herd Groyne lighthouse was built in 1882. It was designed and built by Trinity House Newcastle as Order 611 by its own engineering works. The building was manned until the introduction of electrification and automation early in the twentieth century. In its original state it was provided with living accommodation comprising a bedroom, storeroom, living room, water closet and a coal house.

Another maritime building type that became a familiar feature on Victorian foreshores was the coaling station. Steamers began to venture beyond inland waterways on to the high seas from the 1840s. But while these early steamers had the advantage of not relying on favourable winds, they were constrained by the amount of coal they could carry, which limited them to routes linked by coaling stations. During the steam age these grew to become large utilitarian warehouses often clad in corrugated iron. One early coaling station (*c.*1854), a long barrel-vaulted structure was built at Cape Town, South Africa by Richard Walker.

Changing trade routes were another catalyst for the construction of corrugated iron buildings. During the 1850s, the Isthmus of Panama (the Panama Canal did not open until 1914) became home to a number of corrugated iron buildings. Transport over the isthmus – initially by mule, and later by rail – was by far the quickest route to the American West Coast, which had been opened up thanks largely to the Californian gold rush; until a rail route between New York and California opened in 1869 the only alternative, was by sea, around Cape Horn. A substantial two-storey dwelling supplied by John Walker in 1853 was one of, no doubt, many corrugated iron buildings erected in Chagres during this period.

In 1855, when the railway opened across the Isthmus, it linked the town of Aspinwall to Panama on the Pacific coast. In an account of the new railway and town, an American observer wrote:

> We come upon a building of corrugated iron in progress of erection, intended for the residence of the British Consul, if he will ever have the courage to live in what is only a great target for all the artillery of heaven. The lightning during the rainy season keeps it in a continual blaze of illumination, and I mourned ... over several prostrate cocoa nut palms, which had been struck down in consequence of their fatal propinquity to the iron-house. (Robert Tomes, *Panama in 1855*, New York, 1855)

It was an alarming, albeit uncommon, criticism of corrugated iron.

In the context of transport corrugated iron is most commonly associated with the expansion of the railways. The early rail termini of Britain were enclosed by vast roofs of the material (see Chapter 1). Subsequent rail networks across the developing world followed suit (see Queensland's railways, pages 67–71). But corrugated iron was not only used for showpiece buildings. As the century progressed, it also proved its worth in the design of more humble structures.

OPPOSITE A rare surviving Pagoda-roofed Halt at Doniford Beach station, England.

ABOVE From the 1870s, Roman-era workings at the Dolaucothi Gold Mine in Carmarthenshire were reopened. The corrugated iron buildings erected from the late nineteenth century until the mine finally closed in the 1930s have been rebuilt by the National Trust.

BELOW LEFT Columbia Road fish market, East London, an open-sided large-span structure funded by philanthropist Angela Burdett-Coutts.

BELOW RIGHT The venue for the Industrial and Agricultural Exhibition at Coimbatore, South India, 1857.

By the end of the nineteenth century, railway building in Britain was confined to the completion of branch and local lines. Traffic volume and income was lower, so railway companies tended to spend less money on country stations. From the early-1890s, two standardized building types that proliferated at small stations and halts on the Great Western Railway (GWR) were the Lamp Hut, small structures situated on the extremities of stations to house oil lamps and their fuel away from the rest of the buildings to minimize fire risk, and the Pagoda-roofed Halt. The halts were multi-purpose structures that provided any number of functions on country lines – waiting room, ticket office, storage space. There were once hundreds across Britain, but today only a handful remain, including a listed pagoda at Denham Golf Club Station in Buckinghamshire, and another at Doniford Beach Halt in West Somerset.

Trade and industry

Corrugated iron was born of a need for commercial storage and many of its earliest applications were for large-scale industrial enterprises, such as gas plants and factories. So it is not surprising that by the middle of the nineteenth century the material was still

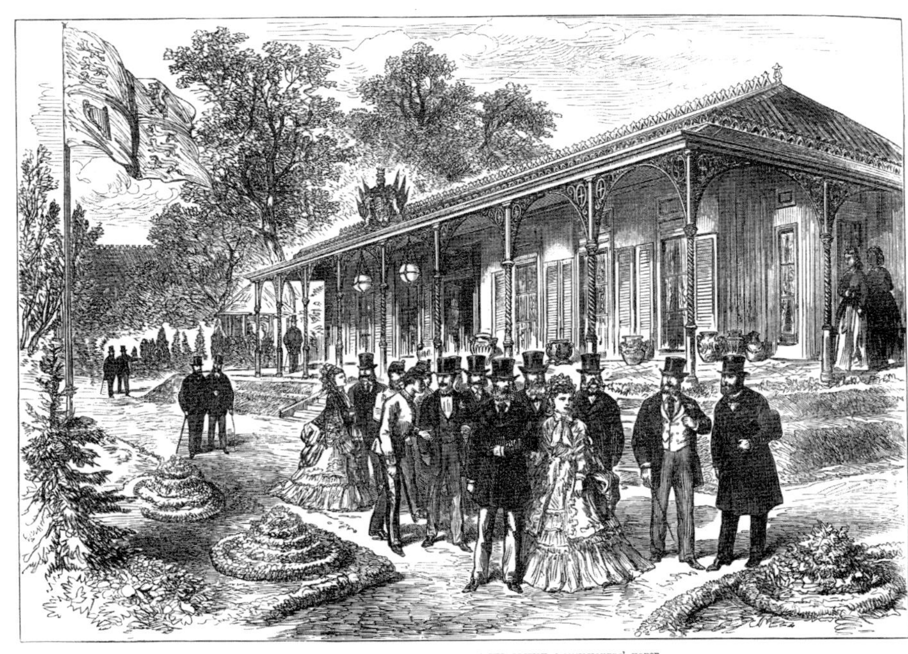
THE VIENNA EXHIBITION: ROYAL PARTY AT THE BRITISH COMMISSIONERS' HOUSE.

LEFT The Prince of Wales, accompanied by the Prince and Princess of Germany, outside the 'elegant little iron pavilion' used by the British Commission to host receptions at the Vienna Exhibition of 1873.

BELOW Tin Town: The temporary town of Birchinlee was built between 1901 and 1903 on a hillside overlooking the site of the Derwent Valley dam. At its height it was home to 900.

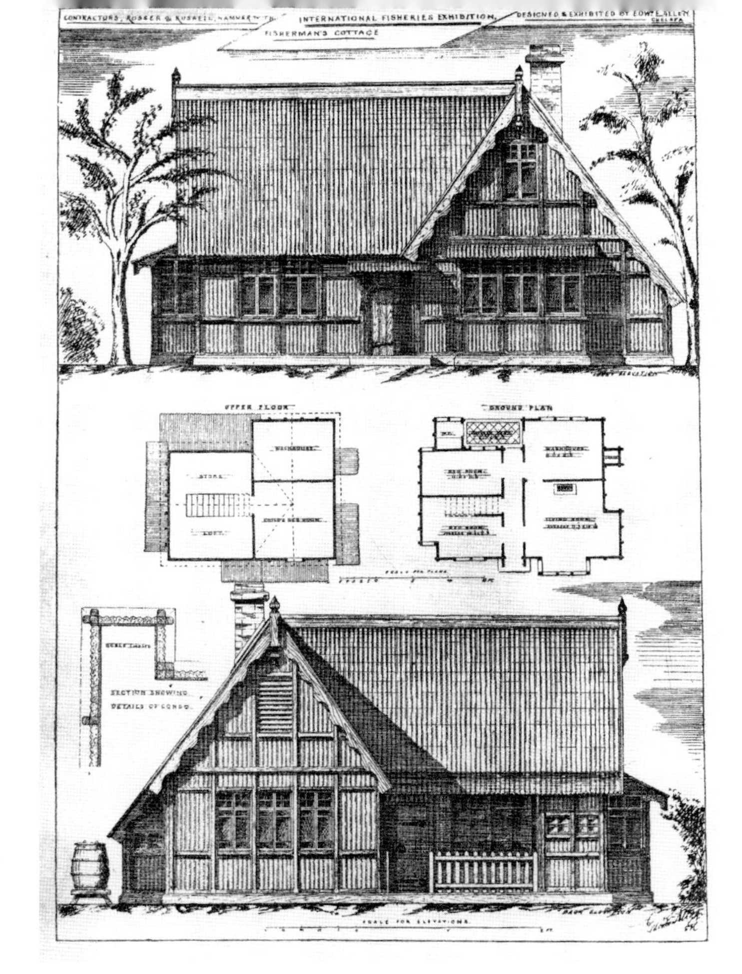

ABOVE Fisherman's Cottage, *c*.1883: To appeal to a more refined market, sheets of iron were fixed into an exposed timber frame, as opposed to overlapping at the edges)

intimately associated with industry; to many that is still the case. But during the mid to late nineteenth century, the material was also used in a remarkably diverse range of applications within the broad spectrum of commerce and industry.

The Customs House that had drawn such large crowds in 1854 prior to its export was one of two buildings designed by E. T. Bellhouse for the town of Paita in Peru (see pages 58–60); the other was a large warehouse. Both were related to Peru's trade in guano and other products. Indeed, Bellhouse generated a great deal of work from South America's growing industrial capacity during the 1850s. He provided a gas works for Buenos Aires, Argentina in 1856 and a 48-sided engine house for the Southern Railway Company in Chile, as well as a large passenger station for the Cantagello Railroad near Rio de Janeiro, Brazil.

Corrugated iron was also commonly used in the design of markets, both enclosed like the market made by John Henderson Porter for San Fernando, Trinidad in 1848, and for open-sided large-span coverings, like the roofs of rail termini. One example of the latter, built in London in the 1870s, was built in the courtyard of the new Columbia Road market in Bethnal Green, East London, a vast development funded by philanthropist Angela Burdett-Coutts (see also, St John's Church, pages 100–3). The main building, an imposing Gothic structure, housed permanent shops and stalls. The covered market provided space for fishmongers and vegetable stallholders. Sadly, the market was an expensive failure and was closed by 1890.

Corrugated iron buildings also became a familiar feature at international trade exhibitions. Examples include the main hall at the Industrial and Agricultural Exhibition at Coimbatore, South India of 1857, and three buildings designed by Samuel Hemming for the Vienna Exhibition of 1873.

As well as two dwelling houses for the 60 workmen employed to maintain the British exhibit, Hemming also designed what the *Illustrated London News* described as an 'elegant little pavilion' enclosing offices, and a large board room for the British Commissioner to host receptions. The pavilion was dressed in Minton tiles and vases, and 'choicest fabrics and upholstery' to impress visiting dignitaries. The overall effect was of a 'snug little nest... [of] bijou... neatness and compactness'.

Samuel Hemming's vision of a whole town built of corrugated iron also became a reality as a consequence of commerce. A Norwegian whaling station on the inhospitable island of South Georgia would not have been commercially viable without it (see pages 72–77).

The town of Birchinlee in Derbyshire was built for different reasons. The Derwent Valley Water Board was set up at the turn of the nineteenth century to build a dam that would flood the Upper Derwent Valley to provide water to the cities of Derby, Leicester, Nottingham and Sheffield. 'Navvies' were recruited for the job – the term was a corruption of the old name for itinerant builders of large civil projects, 'navigators', especially canal builders. At the time, there was widespread criticism of the conditions under which construction workers and their families were living, and because the scheme was miles from a town large enough to accommodate them, the Water Board decided to build a temporary town instead.

A number of companies were asked to quote in late 1900 for the erection of the first fourteen workmen's huts. Messrs Catto, Mather and Co. got the job for £2,143, the lowest tender, for which they would provide huts clad in corrugated iron. In all, by April 1903, around 94 separate buildings were built, all clad in corrugated iron. As well as houses and stores, they included, a hospital, canteen, recreation hall, bathhouse, school, police station and lock-up and a doss house. At its height the town sustained a population of around 900, but once the dam was complete in 1914, Birchinlee was abandoned.

The two-storey tower of Norton Bavant, Hampshire, a corrugated iron house built in the early-1900s, prior to renovation.

Domestic buildings

Many people assume that corrugated iron residential dwellings were temporary, and only prospered in dry, temperate climates, but that overlooks the survival of houses in climates as cold as Iceland and the Falkland Islands, as wet as Scotland and as hot as 'outback' Australia. Another assumption is that corrugated iron cottages were sold only to colonists, or used as makeshift dwellings for the poor, but manufacturers also tried to diversify their markets. During the 1880s corrugated iron 'sporting lodges' were designed with wealthier clients in mind.

One attempt to appeal to the gentry was a Fisherman's Cottage designed by Edward E. Allen, C.E. of Chelsea, and shown at the International Fisheries Exhibition, London of 1883. The cottage was described in *Building News*, 1883 as a, 'new departure in the construction of buildings formed of corrugated galvanised iron', because the 'unsightly' joints between the iron sheets were instead concealed by a timber framework of boards evoking stylistically acceptable half-timbered construction. 'Thus, no laps or joints are visible in any part of the structure except the roof... the effect [has] a pleasing and decided Swiss character.'

Another sporting lodge, designed by Messrs Isaac Dixon and Co. of Hatton Gardens and unveiled at the Liverpool International Exhibition in 1886, took another approach altogether. Very little effort was made to beautify the exterior; according to the *British Architect*, 12 August 1887, the corrugated iron was painted a, 'stone colour ... to [soften] the tone of the whole'. However, the interior was furnished and finished to the highest standards, even including electric lighting.

The survival of nineteenth-century corrugated iron cottages often comes as a surprise. But with maintenance, care and attention, the material has proved itself as resilient as any other, in a range of climates. Hóll Cottage in Reykjavík was built in 1895. The simple three-bedroom, two-storey building still stands; it has been restored as a holiday rental house.

Corrugated iron was introduced to Iceland in the late-1860s. Ships travelling north from Britain to buy sheep would carry cargoes of corrugated iron to sell in Reykjavík, where it quickly became clear that the material was well suited to the isolated volcanic island with limited local construction materials. It is thought that a British-run sulphur mine at Krysuvik was the first building with a corrugated iron roof, but things really took off after 1874 when Reykjavik joined the Fire Insurance Union of Danish Municipalities, which required all new buildings to have fireproof roofs. At the time, only corrugated iron and slate tiles were recognized as fireproof, and as slate was twice the price of iron, the rise of the new material was rapid.

In 1894, local builder ValgarourO Breiofjöro wrote an article for the *Reykvíkingur* newspaper, explaining how to use corrugated iron in the construction of houses. His recommendations were extremely detailed: 'Six nails should be used to fasten each seven-foot sheet of iron to the roof.' He advocated the use of galvanized nails – anything else being a 'false economy' – and said that six nails per sheet was a 'perfect match'. He also described how to prepare sheets for painting: 'First the iron needs to be washed with salt water and left for 14 days... to take the shiniest layer off the iron. Then wash the salt water away... only then is the iron ready to be painted.'

Norton Bavant, a five-bedroom single-storey house in the English village of Beech in Hampshire, is another century-old survivor. It was built in 1903, probably by an officer returning from the Second Boer War. At the time, one way of assisting traumatized young men to find their feet on 'civvy street' was

SEPARATE CATALOGUES.—No. 40, of Iron Churches, Chapels and Schools. No 42, Iron Roofs for all purposes. No. 43, Strained Wire Fencing, Continuous Fencing, Hurdles, Gates, &c., can be had on Application.

DRAWING No. 245.

SUPERIOR IRON RESIDENCE FOR TROPICAL COUNTRIES.

This design is more costly than any of the preceding, and is intended for situations where first cost is not an important consideration. There are several **special features in the design and construction** of this Building which will be fully explained on giving estimate, which can be had on receipt of enquiry with exact particulars of requirements.

☞ **FOR EXPORT these Buildings are fitted together before shipment, carefully marked for re-erection, and we supply corresponding plans, with the assistance of which the erection abroad is readily accomplished.**

⁂ **SPECIAL CARE is taken in Foreign Work to make the constructional details as simple as possible, to facilitate erection abroad.**

EXACT ESTIMATES on receipt of inquiry,—with exact dimensions.

Pages from Isaac Dixon & Co's 1885 catalogue 'Of Improved Iron Buildings for all purposes'.

to grant them small plots of land, often in cheap agricultural areas. Beech, a short distance from the garrison town of Aldershot, was typical.

Until the 1930s, much of the village was built of corrugated iron buildings, including the post office. Today, only two buildings remain, the church of St Peter and 'Norton Bavant' – named after a Wiltshire village from which its original owner may have hailed.

The original part of the house dates from 1903. A two-storey tower and drawing room were added in 1909, giving it a very unusual appearance. For over 60 years, until 1997, it was home to the same occupant, who made very few alterations. When its new owners bought it, they found that the floorboards and wooden frame were largely rotted – because of poor air circulation under the building. But the corrugated iron itself was in generally excellent condition. The Grade II listed house has now been completely rebuilt, retaining its original proportions, and as many features as possible.

It had long been standard practice for corrugated iron manufacturers to promote their products and services in 'small ads' in the trade and technical press. Some also produced short pamphlets, describing their areas of expertise and product range. From the 1870s, as a reflection of the growing scale and stature of the corrugated iron industry, these pamphlets developed into substantial and increasingly luxurious illustrated catalogues.

The catalogue produced by Francis Morton & Co. of Liverpool in 1873, demonstrated both the diversity of the company's range, and documented its earlier successes. Morton & Co. was particularly active in South and Central America. In the 1860s, it supplied a diverse group of buildings to the Cuñapirú Gold Mining Company in Uruguay, including a barrel-topped machinery and stamping shed, a pitched roofed hospital, a workers barracks with veranda and dining hall and a residence for the engineers. Iron Stores and Dwelling Houses, a once common building type in which ground floor commercial premises were surmounted by private accommodation, were sent to Mexico and Turk's Island in the Caribbean. According to Morton's promotional material, the latter was one of the only buildings in the area to survive the, 'fearful tornado of 1866'. Other buildings were designed for use on sugar, coffee and indigo plantations in India and other British colonies. Morton's properties for warmer climates generally featured a veranda,

Continuous Fencing, Hurdles, Gates, &c., can be had on Application

DRAWING No. 252.

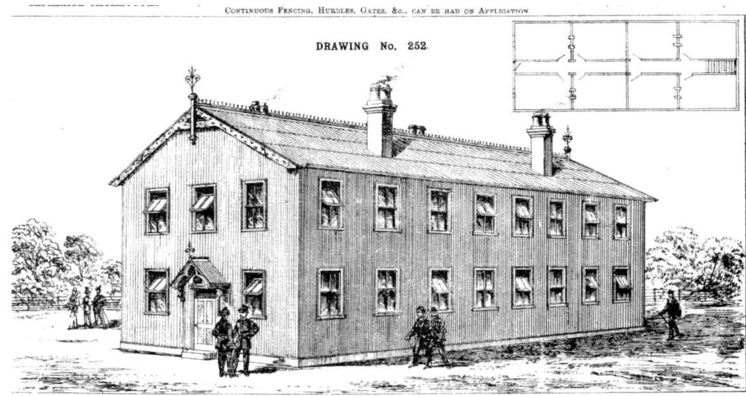

Design of IRON BUILDING suitable for

MILITARY OR POLICE BARRACKS, WORKMEN'S HOUSES, HOSPITALS, &c.

This Building is designed to accommodate a large number of men, or families, each having their own separate rooms, in the simplest form, and at the minimum of cost; such an arrangement affords facilities for heating and lighting the Building. The cost of taking down, removal, and re-erection, in proportion to the extent of accommodation, is reduced to a minimum.

We usually line these buildings with matched boards, stained and varnished, but in some cases the lining of the exterior walls is now dispensed with, and each occupier makes his walls comfortable by means of Felt, Canvas, and Paper, or in a variety of other ways. This design of Building is well worth careful attention, when it is desired to comfortably house a large number of men or families at a moderate cost.

For **Mining, Railway, Manufacturing**, and **Dock Works**, either in this country or abroad, this style of Building is particularly useful. For **Hot Climates** *the Roof can be extended to overhang the walls a few feet and form a verandah all round, which shades the Building, and gives additional coolness and comfort, and also improves the appearance.*

PRICE from about **5s. 6d. to 7s. 6d.** per square foot of ground covered. **EXACT ESTIMATES** on receipt of inquiry,—with exact dimensions.

⁂ **See introductory letterpress giving a few of the advantages peculiar to our Iron Buildings.**

small window openings and substantial ventilation features, while the two supplied to the Right Honourable Lord Hill for erection on the Isle of Skye as guest accommodation and a shooting lodge were of a much more ebullient design, had much larger full-height windows and were supplied with a chimney and fireplace.

Fellow Liverpool manufacturer Isaac Dixon & Co. was just as prolific and had an equally diverse range of products, including skating rinks, sanatoria, entire farms and multi-purpose barracks. William Cooper Ltd, 'Horticultural Providers' of the Old Kent Road in London, was another with a broad range of products by no means exclusive to horticulture. Its catalogues include village halls, isolation hospitals, 'African Merchant's Stations' and a wide range of cottages and bungalows. One of Cooper's 'bungalows', a two-storey house with four bedrooms, still stands in Sherborne, a small town in Dorset, England. Devan Haye was built in 1889, for a John Hall Dalwood, a prominent Methodist, who also had a number of property interests in the area, and it has survived in a remarkably authentic state. A generous hallway is lined on either side with large living spaces, whose internal partition walls remain easy to reposition, depending on need. On the upper level the house

ABOVE Devan Haye in Sherborne, England, a William Cooper-designed 'bungalow' built in 1889.

BELOW Sporting pavilions from John Lysaght Ltd's catalogue of 1910.

OPPOSITE A page from a Boulton & Paul catalogue of around 1900.

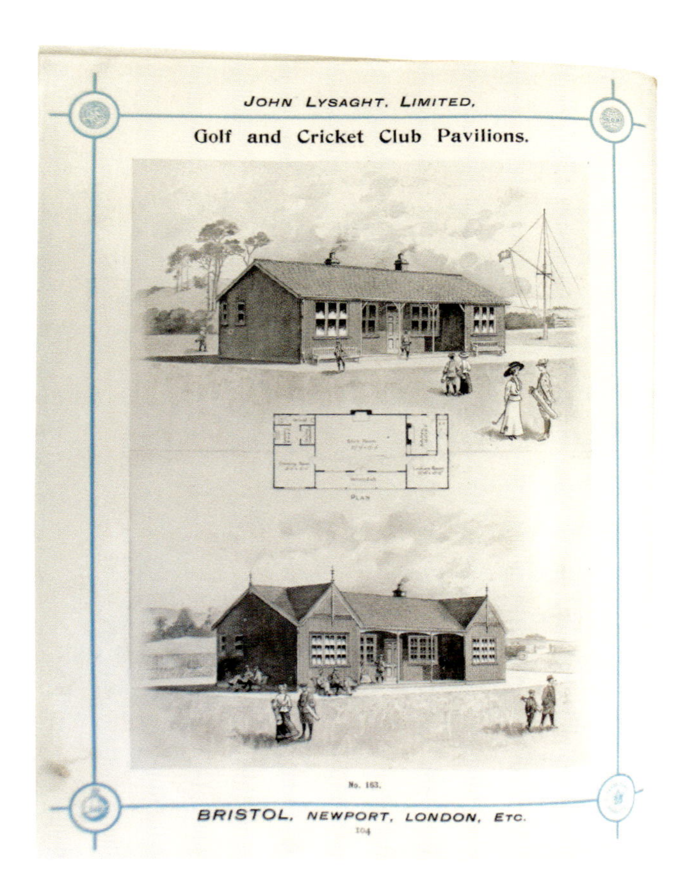

is defined by an enclosed balcony around three sides, creating the impression that the house was designed for a tropical climate, rather than the damp conditions of south west England. Cooper, like most large-scale manufacturers, offered two prices for its products. One was for delivery, 'FOB' (Free On Board) to the nearest 'rail or wharf'; the other included the erection of the building 'on purchaser's foundations'. In the case of 'Devan Haye' the cost differential was £100–£250 for delivery, or £350 for delivery and erection. However, neither price included 'Brick Chimneys, w.c, Fittings [or] Stoves'. Manufacturers at the more exclusive end of the market would have supplied some, if not all of the above.

Boulton & Paul, of Norwich, another heavyweight of the era, fell into that category. The company grew rapidly from the 1870s, thanks largely to its pioneering woven wire fence business, which found a massive market in Australia, where there was a substantial demand for rabbit-proof fences. By the 1880s Boulton & Paul had diversified into a remarkably broad range of iron buildings, many of them corrugated. Catalogues included churches, chapels, parish and village halls and mission rooms (see Chapter 3). Its range also included buildings for the estates of the wealthy, such as gymnasia, boathouses (see title page), shooting boxes and hunting lodges, billiard and smoking rooms, studios and laundries. In the category of municipal and public buildings, Boulton & Paul supplied schools of various sizes, isolation hospitals, village clubhouses, public shelters and cricket, tennis and golf pavilions and, for the farm, its products were legion, including storerooms, shepherd's lambing huts, stables, piggeries, coach houses, cart sheds, game larders, a dairy and numerous other mundane buildings and roofs, all made from

No. 423. GALVANIZED IRON GYMNASIUM,

with Porch, Lavatory, and Visitors' Gallery.

REGISTERED DESIGN, No. 5989.

Constructed according to the Specification on page 92.

TESTIMONIALS.

From THE HON.
HARBORD HARBORD,
Gunton Park.

I beg to inform you that the Iron House built for me suits admirably for what it is adapted, and very easily moved without detriment to the structure. It is perfect and durable, with the additional advantage of its cost being about half of those usually built as a fixture with bricks and mortar, &c.

From
C. J. PRESCOTT, ESQ.,
Wirksworth.

I beg to hand you my cheque for Iron Building which is now erected, and looks very well. If you have any enquiries from this part you can refer them to me, and I will show them your work; they could not have better.

REGISTERED DESIGN, No. 5989.

Interior of No. 423.

TESTIMONIALS.

From
SWAINTON ADAMSON,
ESQ., Rugeley.

GENTLEMEN,—I am very pleased with the House; being first fixed on a good foundation, it makes an excellent house.

From
GILBERT W. STRACEY,
ESQ., Rackheath Park.

SIRS,—I enclose cheque due to you. Sir Henry Stracey is much pleased with the Riding School.

From
MESSRS. JONES & SON,
Denbigh.

We are very much pleased with the structure, and shall have very great pleasure in recommending any of our friends to your Firm.

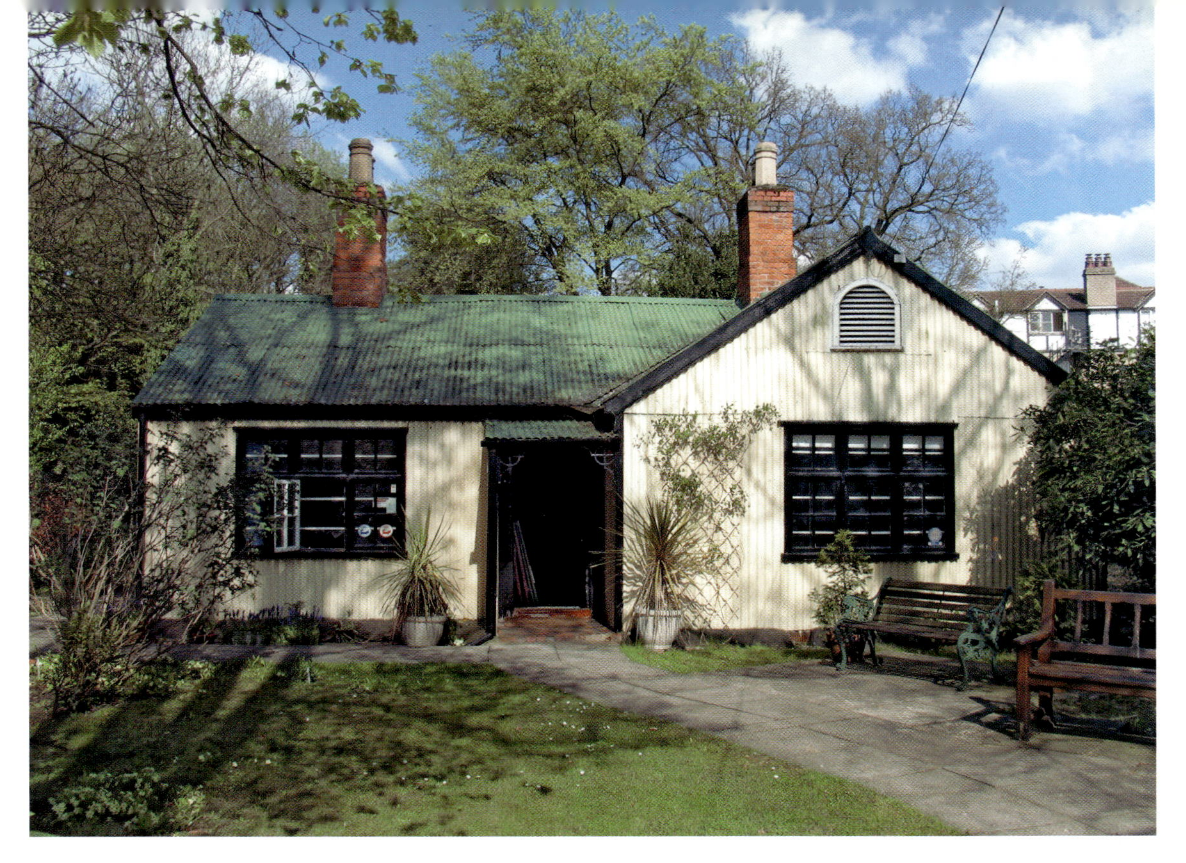

PREVIOUS PAGE A former farmhouse in the Claerwen Valley, Rhayader, Wales.

LEFT In 1887 a corrugated iron cottage selected from the catalogue of Boulton & Paul became home to Thomas and Mary Wield, employees of Lincolnshire's only spa. A century later the cottage became the Woodhall Spa Cottage Museum.

BELOW TOP Excerpt from a Boulton & Paul catalogue, *c.*1890s.

BELOW BOTTOM A corrugated iron smoke house at the Highland Folk Museum, Scotland.

corrugated iron and available to be shipped to the nearest railway station. But it was its range of iron bungalows, cottages and houses for the tropics, leased land and tied labour for which Boulton & Paul are, perhaps, best known. One still stands in the Lincolnshire town of Woodhall Spa.

After noting that his cattle seemed to thrive after drinking water seeping from an abandoned coal shaft, the local squire, Thomas Hotchkiss began to use it to soothe his gouty leg. Word spread and a pump house and brick-lined bath were built in 1830 to treat the patients – 20 to 30 of whom were visiting daily by 1841. The popularity of Lincolnshire's only spa grew through the nineteenth century and facilities were gradually expanded. In 1880 Thomas Wield and his wife, Mary, began work there drawing water and administering the women's baths. In 1887 they were given a corrugated iron clad cottage, 'The Bungalow', as a home and, from this site, Thomas started to make donkey drawn bath chairs to provide a transport service for invalids around the town. The house was manufactured by Boulton & Paul and is the 'No. 415: Galvanized Corrugated Iron Residence' advertised in catalogues of the period. Initially, there were no main services to the building. Water was pumped by hand from an underground catchment, lighting was provided by candles and outside earth closets were used. The Woodhall Spa Cottage Museum opened to the public in May 1987 in The Bungalow as a community museum run by the Woodhall Spa Cottage Museum Trust.

Frederick Braby & Co. Limited of Glasgow was another with a huge output and a wide range of corrugated iron buildings. Amongst its offerings in the 1880s and 1890s were numerous dwellings and stores specifically aimed at the South African and Australian markets; the company also serviced farmers. Braby's portable huts were particularly popular. The company was also one of the largest manufacturers of galvanized corrugated iron sheets, supplying Boulton & Paul, and others, with its raw material.

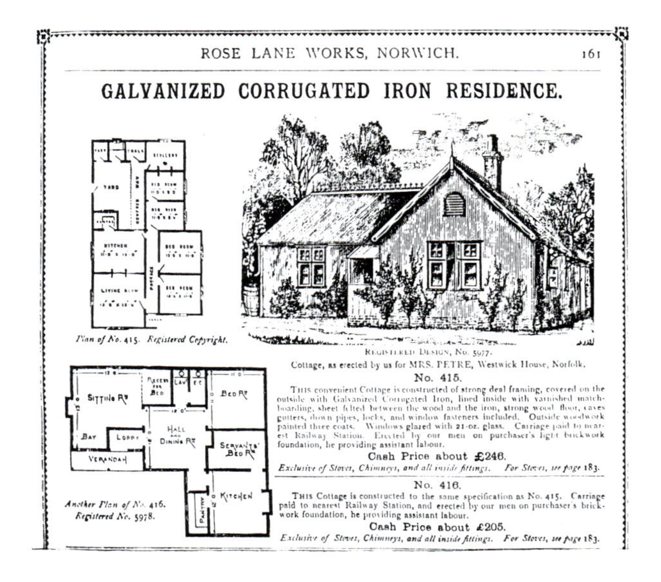

ROSE LANE WORKS, NORWICH. 161

GALVANIZED CORRUGATED IRON RESIDENCE.

Plan of No. 415. Registered Copyright.

REGISTERED DESIGN, No. 5977.

Cottage, as erected by us for MRS. PETRE, Westwick House, Norfolk.

No. 415.

THIS convenient Cottage is constructed of strong deal framing, covered on the outside with Galvanized Corrugated Iron, lined inside with varnished match-boarding, sheet felted between the wood and the iron, strong wood floor, eaves gutters, down pipes, locks, and window fasteners included. Outside woodwork painted three coats. Windows glazed with 21-oz. glass. Carriage paid to nearest Railway Station. Erected by our men on purchaser's light brickwork foundation, he providing assistant labour.

Cash Price about £246.

Exclusive of Stoves, Chimneys, and all inside fittings. For Stoves, see page 183.

Another Plan of No. 416. Registered No. 5978.

No. 416.

THIS Cottage is constructed to the same specification as No. 415. Carriage paid to nearest Railway Station, and erected by our men on purchaser's brick-work foundation, he providing assistant labour.

Cash Price about £205.

Exclusive of Stoves, Chimneys, and all inside fittings. For Stoves, see page 183.

ABOVE Torranbeag Cottage at Ford in Scotland was supplied by Speirs & Co., a Glaswegian manufacturer best known for its churches in the Highlands and Islands.

BELOW Agricultural products available from the catalogue of Hill and Smith Ltd, *c.*1900

FOLLOWING PAGE A Hill and Smith advertisement produced for the Birmingham Cattle Show Guide of 1900.

There were many other manufacturers of conventional (and more unusual) buildings in business as the century came to a close and the twentieth century began. It is not known who built the specialized herring smokehouse, formerly on the Fairburn Estate near Inverness, in around 1900, but the history of Torranbeag Cottage at Ford on the shore of Loch Awe is well documented. It was commissioned from Speirs & Co., a small Glasgow-based manufacturer by the owners of the Loch Awe ferry to house the captain of their new tourist boat. Completed in 1911, it cost £120 including delivery to nearby Ardrishhaig and enough red oxide paint to colour the roof and a stone shade for the walls. It measures about 30 feet long by 28 feet wide and still stands.

Prior to the invention of corrugated iron, buildings were typically made of local materials. But by the outbreak of the First World War, it was a feature of rural and agricultural landscapes throughout the world. It had also played a major role in the religious revivals (see Chapter 3), influenced the practice of farming, provided shelter – temporary and permanent – to millions and was just about to make its marks on the face of modern warfare (Chapter 4). By 1914 corrugated iron had become a truly global construction material.

GALVANIZED STEEL GRANARY.

Design No. 568.

The above illustration shews a Steel Granary on Steel Columns with vermin-proof Caps, which can be made any size to suit purchasers' requirements.

IMPROVED STEEL HAY AND GRAIN SHEDS.

With T Steel Principals having curved Elevated Ties, Angle Iron Eaves Beams and Patent Roof Fasteners.

Design No. 637.

This Shed is very similar to No. 545, as described on back cover, but it is **stronger** and is fitted with the **latest improvements.** It has a **higher rise** to the roof, and the centre tie bars of the principals are curved upwards, so as to give **more headway,** and are connected to the crown of the roof by vertical ties as shewn, which **increase materially the strength** of the structure. HILL & SMITH LTD. have erected many hundreds of this pattern Shed, and they respectfully submit that there is **no better Shed** of its kind **in the market.**

Special Prices quoted upon receipt of requirements.

78

Sole Agents in Great Britain for PORTER'S

For dealing expeditiously with hay—undoubtedly

Design No. 720. Hay Shed.—Fitted with patent Steel Track
Should be usec

Hill & Smith, Ltd., will be pleased to quote special prices, includin;

ATENT STEEL TRACK AND CARRIERS.

best and most reliable system on the market.

Hay Carrier, effecting great economy in time and labour.
every farm.

ing complete, on receipt of particulars as to space to be covered.

BROWN BROTHERS IRON STORE

Mercer and Ginn Streets, Geelong, Victoria, Australia

A heavy masonry facade at ground level was inserted in the early years of the twentieth century, causing severe damage to the foundations.

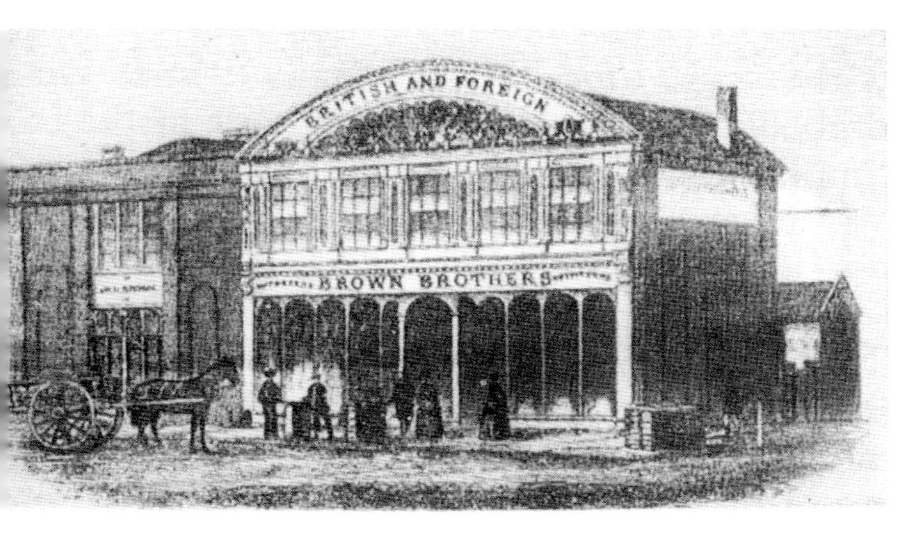

ABOVE LEFT The Iron Store was one of 198 prefabricated buildings that arrived in Geelong in 1854.

ABOVE RIGHT Corrugated iron panels run horizontally above and below the cast-iron window frames, and vertically in between.

LEFT The original Mercer Street elevation had a glazed shop front at ground level, five wood-framed windows above and openwork timber consoles enclosed within the arch of the roof.

The lives of Warren Hastings Brown and his brother George Bogle Brown were shaped by the expansion of the British Empire. The family was well connected and upwardly mobile – Warren Hastings, their father's first cousin, was the first Governor-General of India.

Little is known of the brothers' early years, but during the 1830s they moved from Glasgow to the Caribbean, where they mixed in elevated circles. In the mid-1840s George married Amelia Grano, the widow of the governor of Dominica. By then economic conditions in the Caribbean were in decline, a situation exacerbated by natural disasters and the end of slavery. So in 1851 the Browns returned to Scotland and began searching for new ventures. Later that year gold was discovered in Victoria and New South Wales. Australia beckoned.

On 16 December 1852 the *Sarah Sands* docked in Melbourne. Warren and his brother-in-law George William Henry Grano were among the passengers. They had been sent by the family to weigh up commercial opportunities. Their search took them to Geelong, a port about 50 miles west of Melbourne and gateway to the goldfields around Ballarat. Traffic through the town was substantial.

Anticipating the prospectors' elevated sartorial aspirations, Warren saw potential in the drapery trade. The plan was dispatched to Glasgow. Warren recommended that George import premises for the business. Permanent buildings were scarce in gold rush-time Geelong. If the brothers could start trading as soon as they landed perhaps they could steal a march on the competition.

Back in Scotland George considered the options. Since the Californian gold rush of 1849 iron merchants and manufacturers had sought to capitalize on the market for portable buildings. Robertson & Lister of Glasgow, 'Smiths, Engineers, Millwrights and Iron Roof Constructors', were one of the few enterprises known to have worked with an architect and their collaboration with the Edinburgh practice of Bell & Miller gave their buildings a rare degree of refinement.

From May 1853 fifteen of Robertson & Lister's 'Iron Buildings for the Colonies' were on show in their yard, prior to export, at 340–346 Parliamentary Road. The buildings were made to the specifications of a well-known commercial agent of the day, Hart of Melbourne, and the exhibition was an opportunity for Robertson & Lister to number the components ready for re-erection and to cultivate interest in their products. *The Civil Engineer and Architect's Journal* described the yard as,'a little colony in itself'.

In July 1853 three hundred guests were invited to a ball in the largest of the buildings, a 'saloon' that weighed 25–30 tonnes and cost £600. No expense was spared. The walls were 'adorned with grand plate glass mirrors and hung with pink drapery and evergreens… the floor was covered in white cloth… gasoliers hung from the roof'.

LEFT TOP The Brown Brothers' logo was stamped on the timber shipping cases.

LEFT BELOW The original timber upper façade, an arrangement of segmental arcs with a radiating fan motif, remains substantially intact, complete with embossed mouldings of hemp rope and pre-cast plaster.

It is not known whether George was a guest at the event, but family records confirm that during the summer of 1853 he purchased a two-storey iron building from Robertson & Lister which had living space above an open-plan ground level intended for commercial use. It was the future family home in Australia.

On 16 November George, his wife, two stepchildren, four of his own children, a servant, a nurse and 33 cases boarded the *Ridderkirk*, a 626-tonne Dutch barque, at Gravesend, Kent. They docked at Port Melbourne three months later.

The gold rush had changed the pace of life in Geelong, and Warren Brown had been swept along by the tide. Before the iron store arrived in April – shipped separately from Glasgow – he had already set up the Brown Brothers' drapery business on Malop Street and was living in Geelong's Newtown District. The Brown Brothers still had plans for their store, but first they needed a site.

Mercer Street today is a relatively sleepy road, running parallel with the bend of the bay into the heart of town. But in the 1850s it was the main road through Geelong. Destined to be the town's future commercial core, it ran adjacent to the Geelong–Melbourne rail line, then under construction. Number 17 (at that time numbers 104–106) was located between the dock and two proposed rail stations. Anyone travelling to Melbourne would have had no option but to pass the site. It was prime real estate.

The brothers acquired Allotment 11, Section 42 of the City of Geelong from William Weire, the town clerk, on 16 September 1854. Six weeks later the iron store was complete. The relocation of the drapery business was promoted in the *Geelong Advertiser* on 2 November:

> G. and W. H. Browne [sic] opposite Western and La Trobe Hotels, Mercer Street, have the honor to inform the public of Ashby, Kildar, Little Scotland and elsewhere that on Saturday 1st, they will open a new establishment with a splendid stock of summer goods, newly imported and for sale at prices to satisfy all classes of customers, whom they most respectfully invite to call to test the truth of the above.

Erection of the iron store would not have been a simple exercise. The building is large, 9.8m wide by 12.2m long and 6.9m high (32 by 40 by 22 ½ feet) to the crown of the barrel-vaulted roof, and unusual. It is a rare example of a two-storey prefabricated building still on its original site and it features a full basement and a partially suspended first floor to allow for a column-free commercial space at ground level.

A framework of rolled wrought-iron tees and angles with larger cast-iron angles at the corners is sheathed with galvanized corrugated iron sheets running vertically in the main panels and horizontally in the spandrels below cast-iron window frames. The self-supporting roof is composed of an arch of five corrugated iron sheets, riveted together where they overlap.

The building is carried on a cast-iron inverted T-section perimeter base plate which sits on the stone walls of the basement. Twenty vertical stanchions are fixed in the base plate. Each of the corner stanchions has a recessed face which features a Gothic arch motif below the top plate and first floor levels.

All floors and internal partitions are timber. Those still marked 'BB', a reference to the Brown Brothers, reveal that the timber shipping cases were used as internal partitions, a common practice. The walls were originally covered in hessian and thick floral wallpaper, much of which has survived.

More than 150 years after it left Glasgow the building is in most respects intact but for the appearance of the façade. An 1859 invoice head shows a glazed shop front framed in ornate cast iron; there were four narrow arches on either side of a central doorway. The design is not included in any of Robertson & Lister's surviving drawings, but one similar – a drawing of an 'Iron Dwelling House and Store with Handsome Cast Iron Front' – was published in *Illustrations of Iron Buildings for Home and Abroad*,

Timber panels from the original shipping cases were recycled as internal partition walls.

an 1856 catalogue produced by Charles D. Young & Co., which was almost certainly associated with Robertson & Lister during the 1850s. One explanation is that George Brown specified a façade appropriate for use in a drapery store – one that would allow light to flood in and passers-by to see the quality of their merchandise – when he bought the building.

The building has endured far longer than the Browns' enterprise; the brothers' partnership was dissolved after only five years. In 1860 the firm of Brown, Osborne & Co., 'forwarding agents', began operating from the premises and the building remained in the ownership of the company until 1872. Over the next thirty years it was occupied by a lime burner, a saddler and a succession of storekeepers.

In October 1909 the building was bought by William Nash, a grocer, for £275. It is thought that he was responsible for the insertion of the masonry façade at ground level. His daughter, Miss J. Nash, lived there until she died in 1956. The Erskine family, who used the building as a depository for military goods, moved in soon afterwards and stayed until 2000. By then it was in an advanced state of dereliction. A 1997 conservation plan revealed serious underlying problems, notably the condition of the front basement wall and the leaking roof. In 2002 archaeologist Brendan Marshall bought the iron store. He plans to restore it to its original state and to use it as the headquarters for his business.

CUSTOMS HOUSE

Paita, Peru

Paita, a port on the Pacific Ocean, is the most northerly settlement in Peru, just five degrees south of the equator. A traveller in 1865 described the town:

> The harbor is very accessible and has a fine depth of water; our ship is anchored in nine fathoms, half a mile from shore. The appearance of Paita is not prepossessing. The houses are clustered closely on the beach at the foot of a sand bluff about one hundred and fifty feet high... A substantial mole [pier] receives passengers on landing, and on this they pass a short distance to the custom house, an iron structure prepared in England for erection on its arrival here a few years since. It is a two-storey building... surrounded by a neat iron balcony, and is surmounted by a cupola, from the staff of which floats the Peruvian flag. Behind the custom house, is the public store, of the same material. All other houses are built in the rudest manner and of indifferent materials; usually rough timber frames, filled in with clumsy sun-dried bricks – adobes of all shapes and sizes; or Guayaquil reeds, whole or split, daubed with mud... The streets are narrow, irregular, unpaved; and when you are informed that the bluff at the rear of the town is the commencement of ... a desert on which rain rarely falls and dews never – you may imagine the depths and drifts of sand blown from the neighbouring hills presenting a picture of barrenness and desolation. (Henry Willis Baxter, *What I saw on the West Coast of South America...*, 1865)

Eleven years earlier the same customs house – a building in which customs officials collect import and export duties – had been proudly flying the Union Jack from its flagstaff 100 feet above the ground in Granby Row, Manchester, where, according to *The Builder* of 4 March 1854, it had been 'put together in a temporary manner, so as to give assurance of its being perfect in all its parts'. The building was in Messrs E. T. Bellhouse and Co.'s yard for ten days, 'attracting considerable attention and interest... by... being visited and inspected by 25,000 persons' as the *Civil Engineer and Architects Journal* noted. At the time, in 1854, it was considered 'one of the most important works in iron ever executed' (*Practical Mechanics Journal*, Vol.VII, 1854).

Bellhouse's Peruvian order was for two buildings. Firstly and most impressively, the 70 square feet, two-storey customs house; secondly: a sizeable warehouse complex with a courtyard measuring 120 by 90 feet.

The customs house was refined, detailed and sensitively proportioned by Edward Salmons, a well-known local architect, who later worked with

BELOW TOP The otherwise unprepossessing town of Paita was adorned with 'one of the most important works in iron ever executed'.

BELOW BOTTOM Prior to export the Customs House attracted 25,000 visitors in ten days to Edward Bellhouse's Manchester works.

OPPOSITE The frame of the Customs House probably still stands; a stucco-clad building of almost exactly the same proportions occupies the same site at the end of the pier.

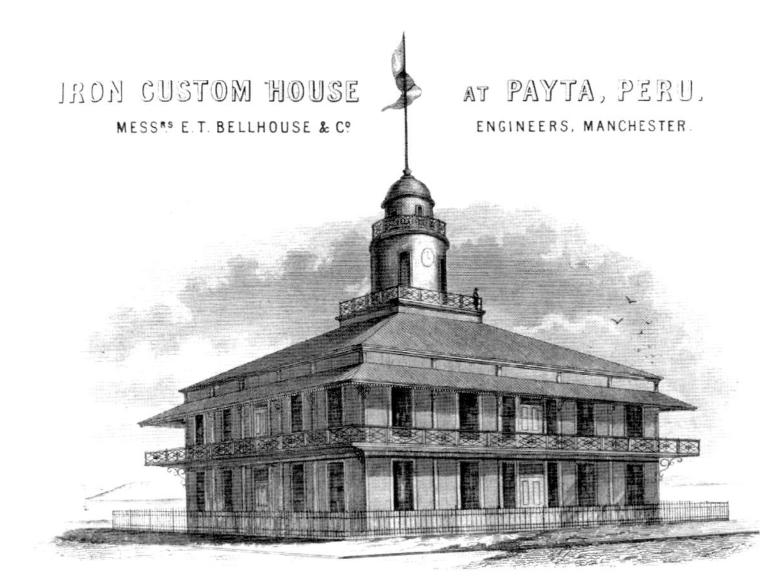

Charles D. Young and Co. on the iron and glass Manchester Art Treasures building in 1856. The walls were of extra bold galvanized iron sheets with corrugations five inches wide, arranged vertically between elegant cast-iron pilasters. These pilasters divided each of the elevations into five bays of equal width. In the central bay of each elevation there was a wide door with sidelights. On either side of these doors two generous French windows gave ample light to the rooms behind. The four elevations of this perfectly square building were identical. On the ground floor an ample 13-feet wide passage ran through the building. Below the cantilever brackets supporting the 6-feet wide verandah that surrounded the building, a carefully proportioned cast-iron cornice crowned the 12-feet high ground floor wall, while the first floor verandah roof was supported on bold cast-iron cantilever brackets trimmed with a decorative valence. The accessible belvedere, 22½ feet square, was surrounded by a light iron balustrade matching that of the verandah at first floor. Above it was the two-level circular tower crowning a pyramidal roof ornamented with cornices and with a very generous loft space. This was topped by a cupola and flagpole. A large public clock completed the composition.

The structure of the customs house was a combination of cast-iron supports and wrought-iron framing. The floors and joists were in timber as were the internal wall linings, which were covered in felt and papered. The cast-iron pillars supporting the central tower were cunningly concealed within the thickness of the partitions.

The design is remarkable for its simple and balanced proportions. It is also extremely efficient at creating a very large built volume with very little material. It clearly stood out in striking contrast to the simple mud buildings that surrounded it in Paita. The visual impact of its clean bold lines would have been powerful at its destination, controlling the land end of a shipping jetty against clear equatorial skies. It would have been even stranger there than in a factory yard in Manchester.

The warehouse complex, by no means unimpressive in size, was built according to Bellhouse's patent system with walls constructed with horizontal corrugated sheets bolted onto specially formed pilasters.

These iron buildings were ordered by the Peruvian Treasury Department. Paita was a busy port near the guano deposits owned by the state and by far the most lucrative export for Peru. Paita was also a major port of call for coastal steamers and for the whaling fleets in the South Pacific, who came regularly for supplies of fresh food and water. All produce and water had to be brought in by mule from quite a considerable distance across the desert, but the anchorage was fine. Its importance can be gauged by the fact that despite the small population of about five thousand, the town had British, French and American consuls. At the end of the nineteenth century Paita provided one third of Peru's customs revenues.

The government had published an official communiqué in *El Comercio*, a leading Lima newspaper, in February 1853 announcing their intention to build a customs house at Paita. The existing buildings were in a ruinous state and the importance of the port warranted a new building. Plans had already been approved by the government in quite some detail with dimensions mentioned that are very similar to the buildings delivered. The site was to be on land formerly occupied by the Convento de la Merced. The iron in the walls and roof was to be of a thickness greater than customarily used in such buildings and it was noted that it should be subjected to the operation of galvanizing to prevent the oxidizing

LEFT The Customs House is remarkable for its simplicity: very little iron was used in the creation of the large volume.

BELOW Paita, a deep water port on the Pacific Ocean, was located close to extensive guano deposits.

effects of sea and air. The supplier was obliged to see to it that the construction was solid while conforming to the best rules of architecture. All materials would enter the country free of import duty. The report ends by scheduling the method and timing of payments which were to be charged against consignments of guano in England at a specified rate of exchange.

A report about the building in the British press states that the execution of these large iron buildings has been intrusted (*sic*) by Messrs A. Gibbs and Sons, London, the banking house that had a monopoly of the guano trade and that they had been constructed under the superintendence of Mr. Edward Woods, C.E. of London, who at the time was a consulting engineer with many railway projects in South America including the Mollenda–Arequipa and Callao–Oroyo lines in Peru.

Arranging for the construction of substantial buildings, with materials, techniques and expertise supplied from across the world, was a complex business. Both sides had to satisfy themselves that their expenditure of time and money was to be rewarded. Many intermediaries were involved acting for governments and other interests, while risks of loss through shipping hazards and unforseen conflicts as well as currency fluctuations had to be taken into account. Buildings were often part of larger civil engineering projects or extended relationships which kept many engineers and manufacturing firms in Britain and other centres of industry busy servicing their lucrative export markets.

Paita's Customs House probably still stands at the end of the pier at Paita. There is a structure whose general shape and size matches the building sent out 150 years ago. The iron is gone as is the verandah, but the original Bellhouse structure may well be hidden inside the current crumbling stucco skin.

THE REAL BODEGA DE LA CONCHA

Jerez de la Frontera, Spain

RIGHT TOP The iron roof is supported by 24 radiating iron girders which meet at a ring that carries the ventilating lantern.

RIGHT BOTTOM The Real Bodega has often been used as a venue to entertain eminent guests. In 1904 it was decorated for the visit of King Alfonso XIII.

The origin and history of the Real Bodega de la Concha, a sherry warehouse with a circular corrugated iron roof, is immersed in myth. Some say it was designed by Gustave Eiffel, others claim that its 'real' (royal) prefix dates to a visit from Queen Isabella II in 1862. Neither is true, although both have drawn attention to the vineyard, and no doubt sales have been enhanced as a consequence.

The grand circular structure was erected on the premises of wine grower and sherry producer González Byass in Jerez, Andalucia during 1868 and 1869. It was commissioned at the height of the González Byass' wealth, and the family wanted a building that would reflect its power. Today it is recognized as a fine and innovative example of vineyard architecture, one that is representative of late nineteenth-century industrial design. It uses rolled and fabricated iron girders and iron in its cladding.

The Bodega de la Concha is free from internal structure that might hinder the movement of large barrels inside it. The roof has 24 wrought-iron lattice girders that rest on a perimeter masonry wall, creating an open space 100 feet in diameter. The lightweight roof of corrugated iron helps to make this large expanse possible. The galvanized sheets were specified to reflect the sun's rays; they performed a similar role to the shiny Moorish tiles traditionally used for the role.

The roof rises to a glazed central ventilator 22 feet in diameter topped by an ornamental vane 12 feet high. Ventilation is provided by a ring of twelve pairs of louvered apertures fabricated into corrugated iron panels set into the sides of the ventilator. Air circulation is also assisted by the orientation of the bodega south west to north east. The open side faces south west.

According to an article in the *Engineer* of 18 March 1870, the building was designed by Joseph Coogan, an Irish civil engineer about whom very little is known, and built by Hermanos Portilla y White, a company that started life as grain millers and later became pioneers in Spain's iron fabrication and mechanical engineering industries.

The story of the bodega's regal prefix is often associated with the visit to Jerez by Queen Isabella II of Spain in 1862. Her Highness did

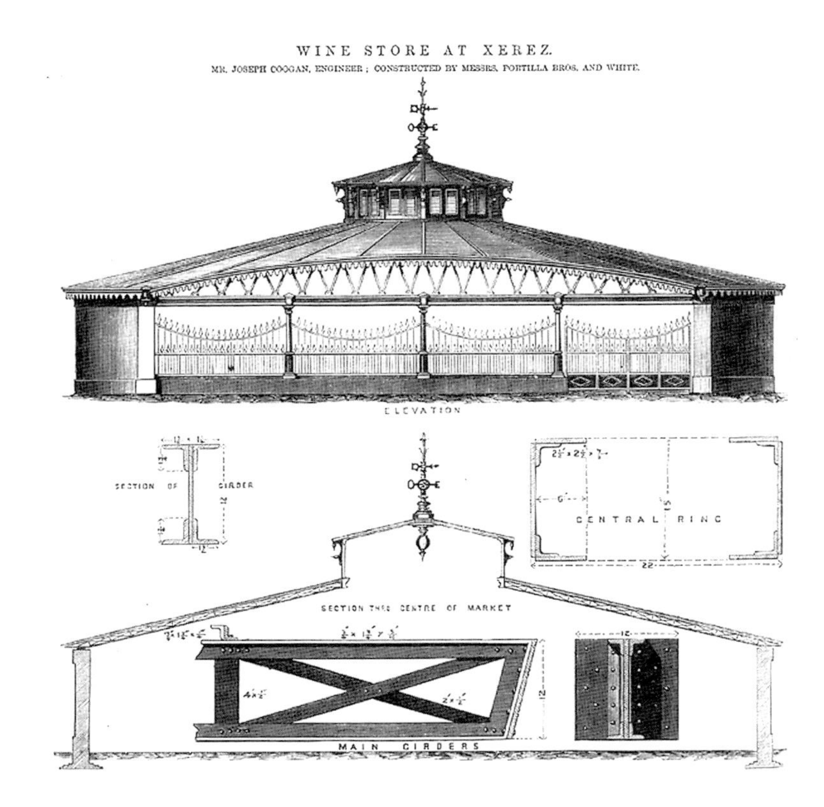

ABOVE The bodega is free from any internal structure that might hinder the movement of barrels.

OPPOSITE The Real Bodega was erected by Hermanos Portilla y White, a leading firm of Spanish fabricators in 1869, to a design by Irish civil engineer Joseph Coogan.

indeed visit in that year, to show support for a new local governor and to oversee provision of potable water to the town. And she did visit González Byass, but she did not visit the corrugated iron warehouse because it had not been built.

The first members of the Spanish royal family, King Alfonso XII and his second wife Queen Maria Christina of Austria, to step inside the building did not do so until 1877, when a banquet was provided in their honour. King Alfonso XIII ate there in 1904 and King Juan Carlos and Queen Sofia were entertained there in 2000.

In its early years the bodega was used to lay up young wines before they were moved for ageing and maturation. Over the years, aside from its role as a venue for entertaining guests, it has been used as a cellar and showroom. Today the building houses 214 casks of 500-litre capacity marked with the flags of the 115 countries to which González Byass exports wines and sherries.

PILGRIMS REST

Mpumalanga, South Africa

BELOW Access to the Transvaal goldfields required wagon transport. From the coast, the track went through swamps and then climbed 6,000 feet over a total distance of 165 miles. Progress was slow, averaging 21 miles per day.

Within a few years of the Californian rush, prospecting for gold and other minerals became an established international enterprise. For the remainder of the nineteenth century there was a steady stream of bonanzas in locations around the world. All of them attracted rabbles of veteran fortune hunters, opportunist adventurers and novices; the possibility of making money without a substantial capital investment was too good to miss.

It was typical for the first wave of gold seekers to live in tents. As time passed, the settlements gradually developed a less makeshift character, with buildings made of local or imported timber and corrugated iron. When the mineral ran out or richer deposits were found elsewhere these temporary settlements often became ghost towns. But some have survived, including the small town of Pilgrims Rest, now in the Province of Mpumalanga in South Africa.

From the earliest years of European trading, it had been well known that Africa had rich gold deposits, but unlike other El Dorados, the interior of the continent was inaccessible to outsiders until well into the second half of the nineteenth century. The almost insurmountable barrier to prospectors was their vulnerability to tropical diseases and the tsetse fly and other pests and parasites capable of killing horses and other beasts of burden.

The South African Republic, established in 1852 by Boers (farmers) of Dutch descent escaping British rule in the Cape Colony, occupied the Transvaal, a large landlocked territory, high enough to be free from tropical diseases affecting people and animals. Gold had been discovered there as early as 1854. But the government of the time, for political reasons and afraid that the country would be overrun by all sorts and conditions of men, had no intention of making use of the discovery and declared that the finding of gold would be punishable with a fine of £500.

This policy was reversed by President Burgers in 1872, who offered rewards for 'payable goldfields'. It was this that led to the opening up of Pilgrims Rest in a remote mountain valley. According to Annie Russell, an early commentator, the settlement was so called, 'because of the rest… suggested when the almost unconquerable task of reaching it had been accomplished'.

ABOVE The buildings of Pilgrims Rest are composed of recycled sheets of corrugated iron and other second-hand building materials. No two buildings are the same.

BELOW The bar of the Royal Hotel began life as a church in Lourenço Marques, a Portuguese port that serviced the Transvaal goldfields.

ABOVE Within two years of the first gold strike, Pilgrims Rest was a flourishing township with a population of up to 800.

BELOW The corrugating machine at Pilgrims Rest would have been used to curve iron sheets for use in water tanks.

In another article she had written:

> The success of a few diggers had the effect of creating some excitement… and in less than two years a flourishing little township numbering a population of four to eight hundred souls had grown up in the picturesque isolation of the mountains. Stores were established, law agents prospered, a bank, temporary places of worship, a newspaper, all found new pastures for their extended efforts… The population of the gold fields consisted altogether of English people.

Legend has it that the first gold in this picturesque valley was found by the eccentric Alec 'Wheelbarrow' Patterson, a prospector who pushed his possessions in a wheelbarrow 1,600 miles across the veld from the Cape.

As in other gold settlements, the first wave of diggers lived in tents, but gradually a more durable town emerged, made up almost entirely of corrugated iron buildings. There was no local timber worth milling, and building in masonry and thatch was time-consuming and burdensome in terms of maintenance.

The corrugated iron buildings at Pilgrims Rest are all individual and unlikely to have been ordered from any of the well-known manufacturers in Britain or elsewhere. Instead the itinerant population used the material's unique properties to create the largest possible internal volume with as little material as possible.

The buildings are minimally framed with imported softwood and lined with thin match boarding. A porter bringing the material from the nearest port at Delagoa Bay, 165 miles away, could carry 70 square feet of covering material for a roof or walls in a single consignment, and everyone knew how to build with corrugated iron. The materials that went into this type of construction had the added advantage of being reusable elsewhere if the diggings did not prosper.

The bar of the Royal Hotel at Pilgrims Rest perfectly illustrates the advantage of portability. It started life as a small Catholic church in Lourenço Marques – the nearest port in Portuguese territory, which prospered by servicing the Transvaal goldfields – before it was dismantled and carried up to its present location.

Pilgrims Rest's life as a centre of alluvial gold mining was short-lived. More profitable mining dependent on capital-intensive machinery followed in the area and later started on a vast scale in the goldfields of the Witwatersrand near Johannesburg. Mining companies sometimes housed their mineworkers in standardized corrugated iron buildings while vast industrial structures clad in the same material housed their stamping mills.

Mining remained active in Pilgrims Rest until 1971. Since then the South African government has declared the town a national monument. It has become a popular tourist destination and open-air museum.

QUEENSLAND'S RAILWAY STATIONS

Queensland, Australia

BELOW Perspective of Toowoomba rail terminus, *c*.1867. Curiously there are few references to traditional railway adjuncts, such as tracks or trains. The inclusion of tropical vegetation and dark-skinned, turbaned men suggests that the illustrator knew little of Queensland's climate or indigenous population.

Toowoomba's enormous terminal station, manufactured in Britain under the supervision of Sir Charles Fox, arrived in Queensland in 1867. But it was completely out of scale for the small inland town and was never erected at its intended site. It also arrived at a time of economic crisis in the young colony, a fact that attracted strong condemnation of the building. Had Queensland been booming, things might have been very different.

It is unclear who imagined that Toowoomba would have needed such a large terminus. Perhaps it was enthusiasm for the railway as 'agent of change' that contributed to unbounded optimism. It may also have been the product of misunderstandings magnified by the long lines of communication. Whatever the explanation, the experience helped to engender a spirit of self-reliance in Queensland. It also established a deep suspicion of buildings imported from the mother country.

An illustration of the terminus, a two-storey corrugated iron building, was published in the 1 June 1867 edition of *The Builder*, alongside an article on the new narrow gauge railway that offered the key to prosperity for the new colony. Toowoomba, separated from the coast by two mountain ranges and difficult terrain, was considered an ideal collecting point for pastoral goods from an extensive hinterland. The railway would allow wool and other products to be exported to markets overseas.

The design was credited to Messrs J. & R. Fisher who were listed in an 1869 prospectus as 'Architects and Engineers' to Frederick Braby & Co. The building was described as suitable for a hot climate with wide verandas to protect the corrugated iron wall cladding and the generously tall windows and doors from the heat of the sun. The published perspective shows the station has a relentless repetitive structure hidden beneath the surface. Conceptually, at least in this respect, there are parallels with the 1851 Great Exhibition building. This is not surprising as Charles Fox was the Crystal Palace's main contractor and his team had turned Paxton's ideas into an achievable reality.

LEFT An illustration of the Ipswich terminus, *c*.1868. It was not as grand as the proposed Toowoomba station, but it was built and remained operational for several decades.

BELOW Rockhampton terminus in 1912. The station is clearly related in design to the Toowoomba terminus albeit on a much more modest scale.

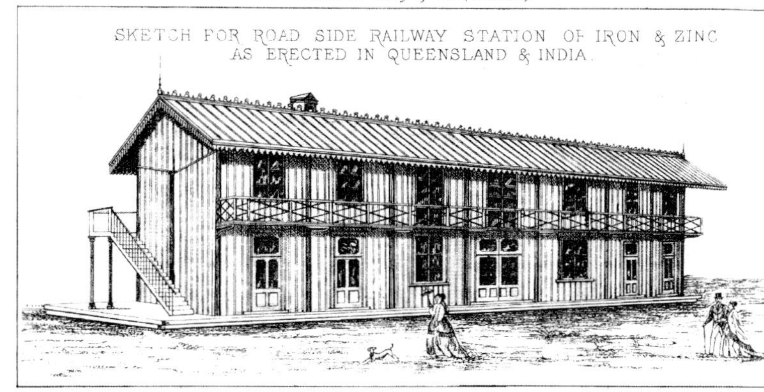

LEFT TOP A wayside station from Frederick Braby & Co.'s 1869 catalogue. The design is attributed to J. & R. Fisher who also designed Toowoomba station. Note the similarity with Laidley station.

LEFT BOTTOM Laidley station, c.1868, on the line between Ipswich and Toowoomba. One of the smaller imported iron buildings on the Southern and Western Railway.

FOLLOWING PAGE The Toowoomba terminus was not entirely lost. The train shed from the original building (behind the masonry building in the picture) was erected in 1874 when the railway was extended from Ipswich to its Brisbane Terminus at Roma Street.

The Builder described the terminus as:

> ... planned in cubes of 15ft, having the columns, standards, girders, &c 15ft apart with a principal [structural element] over each. The external walls consist of cast-iron H standards, vertical and horizontal wrought-iron girders, with horizontal wood transoms, to which are attached the outside corrugated iron and inside wood lining, leaving a space between for a current of air.

The free-standing ornamental veranda columns and lattice girders are the visible part of this modular idea. In this building, however, the consequences of modularization impinge less rigidly than in the many other such buildings built at the time. Accepting assembly as a necessity without the need to accentuate it at the expense of mass, form and an air of permanence would have been seen as beneficial to the aesthetic sensibilities of the time. In addition, efforts were made to introduce variety, emphasis and incident to add interest to a pragmatically conceived construction. Readers of *The Builder* were, however, reassured of the practicality and economy of this large prefabricated project:

> The whole was put together in England, so as to save trouble and expense in the colony. The total cost, including all fittings, was 6d. per cubic foot, equal to about 8d. when erected in the colony.

The station arrived in Queensland at a very bad time. The colony was in dire financial straits, and the railway was singled out for meticulous scrutiny in the Queensland parliament. In 1869 William Mason, the Acting Engineer-in-Chief for Railways in New South Wales, reported to the Parliamentary enquiry:

> Amongst the numerous and palpable errors which have been committed in the designing and construction of these railways, none stands out more prominently than the station buildings – errors which cannot fail to strike the most casual observer... [Toowoomba station], which is entirely of iron, was ordered and delivered in the colony, but not erected, and is now laid by as useless. Judging from the drawings, it is both in size and appearance more like a palace than an ordinary station building; and, when erected, would have occupied nearly five-eighths of an acre of ground! ... It would be a superfluous remark to say that such a station as that originally designed, ordered, and delivered, was unnecessary for such a place as Toowoomba, which contains only 2,901 inhabitants. To seriously propose such a thing would, in itself, be a gross outrage of common sense; but to actually order it, receive it, and then cast it aside as useless and waste, is an unprecedented and unjustified piece of extravagance, which warrants the severest condemnation. That such a building was unnecessary, is not only an opinion, it is a fact already proved by the erection of the present station, which, as I have already mentioned, has been found sufficient for the traffic, and likely to be so for several years... Altogether, it is a great and serious blunder.

Toowomba Station was not the only building to have been sent with all the railway bridges, permanent wayside stations, equipment and rolling stock for the two Queensland railways being built at the time. In addition to extensive workshops and some smaller wayside stations, two other substantial termini, also by Braby, albeit on a smaller scale were erected without much fuss. It was argued, again in *The Builder*, that they were: '... designed to meet the requirements of a country where materials are difficult to be obtained, and where facility of erection is of importance'.

The terminus at Ipswich where the railway could tranship goods to shipping on the Brisbane River became the nucleus of extensive railway workshops that attracted other engineering industries to the town. The building was not as grand as that intended for Toowoomba, but survived for several decades. The third terminus, in a similar style but smaller than the Toowoomba building, was erected at Rockhampton on another railway line in Queensland built at the same time. It continued in service well into the twentieth century.

Braby also supplied smaller wayside stations, including one erected at Laidley on the Southern and Western Railway. They hoped to stimulate a wider interest in their ability to make station buildings by including a drawing of a similar structure in their 1869 catalogue.

When the railway was extended from Ipswich to Brisbane in 1874, part of the Toowoomba terminus was erected as an adjunct to a conventional brick building at Roma Street station. The large shed, covering the railway tracks (carefully concealed in the view published in *The Builder*) was erected at this station.

This collection of termini and other railway buildings was almost certainly the largest single group of prefabricated structures sent to Australia in the nineteenth century. They were all clad entirely in corrugated iron.

GRYTVIKEN WHALING STATION

South Georgia

Without corrugated iron kit buildings, life for the first fleet of Norwegian whalers in South Georgia would have been impossible. The island is barren, remote and battered by freezing winds. It is certainly no place for human habitation. But the Norwegians knew that there would be substantial financial gain in the endeavour.

Captain Karl Anton Larson led the whalers. It was his fourth trip to the island. His previous visits had taught him much about the dangers and riches to be found there. In 1901 he was commissioned to captain the *Antarctic* during an expedition to the South Atlantic led by Dr Otto Nordenskjöld, a Swedish geologist.

The expedition left Gothenburg in October 1901 and reached South Georgia the following January. While visiting the north of the island, Larsen and his companions found two huge three-legged iron cauldrons abandoned by sealers in Cumberland Bay East – the island's fur and elephant seal populations were hunted almost to the point of extinction during the first half of the nineteenth century. They named the spot Grytviken (Cauldron Harbour). It was further evidence of the natural resources on the island. But Larsen nearly perished before he could exploit them. On its return leg the *Antarctic* was crushed by pack ice in the Weddell Sea. The rescue of all members of the party by an Argentinian corvette has become the stuff of legend.

While recovering in Buenos Aires, Larsen formed the Compañia Argentina de Pesca, with mainly Argentine financial assistance, to hunt for whales in the seas around South Georgia. Whaling was extremely lucrative at the time, with products used for heating and lubrication, and in the production of soap, margarine and baleen (whalebone) for corsets.

He also negotiated a licence from the British government – South Georgia had been a British territory since Captain James Cook set foot on the island in January 1775.

Larsen sourced much of the equipment needed to establish the whaling station from Norway. In his hometown of Sandefjord, he contacted an old acquaintance Ole Wegger, who was a director of the shipbuilding company A/S Framnaes Mekaniske Værksted, from which he ordered his whale-catcher, the *Fortuna*. The company also provided him with industrial plant and equipment and residential and factory buildings, two of which may have been subcontracted from Strømmen Trævarefabrik.

This company was originally established in 1884 by engineers Christen Arena Segelke and Gabriel Kielland Drift to supply timber building components. By 1900 they were making complete structures, including workmen's cottages, villas and churches. The buildings were often designed in conjunction with architects, and most followed a traditional Scandinavian pattern with wide eaves to deflect winter snows. Many were produced with wooden frames and corrugated iron cladding, including those ordered for Grytviken.

Sunday afternoon football, *c.*1920: At its peak, Grytviken included foundries, blacksmiths, laboratories, a hospital, a hydroelectric power station, a church, a cinema and a ski jump for recreation.

One of the buildings chosen by Larsen was to house the factory equipment, which consisted of twelve open cookers for melting blubber, two rubber presses, two steam boilers and the cooper's workshop. The other was the workers' dormitory; known as Barraca No. 1 on Larsen's 1911 plan of the settlement, and later the Russebrakke (or Russian House), where Russian and Slav workers were billeted. The accommodation block still stands. Both, including their contents, cost less than 8,500 kroner (around £450).

Before the ship set sail, the buildings were erected on the dock at Sandefjord by some of the workers recruited by Karl Larsen for the new station. They were then taken down again and packed into one of the

A sperm whale being flensed, *c.*1920s.

other two ships procured for the small fleet. The party left for the South Atlantic on 26 July 1904.

The first two ships arrived in Grytviken on 16 November followed on 28 December by the third member of the fleet carrying a small wooden bungalow, which became the administrative office and Larsen's home on South Georgia. All three buildings were erected by the community of around 40 men within five weeks of their arrival. As soon as these were completed, the men started work on a further six corrugated iron workshops, stores and farm sheds, all of which were completed before the end of the southern summer.

The whaling operations were an immediate success. In the first season 183 whales were killed, producing 5,697 barrels of oil. Within the first few years investors were being paid 50 per cent dividends. By 1912 six more stations had been established on the island. At its peak Grytviken had a population of 500 men.

The Grytviken station operated for 61 years catching blue, fin, sei and sperm whales. Over 30,000 were killed and processed in and around the island each year during the 1930s alone.

Every spring transport ships brought the materials required for the summer season's whaling, rendering the small community virtually self-sufficient. The boats brought provisions for the men and coal (later fuel oil) to fire the steam boilers. These boilers powered bone saws, hauling winches, blubber boilers, skinners and drive shafts for the machine shops. When in full operation the bay was filled with clouds of foul smelling vapour and smoke.

Pigs and chickens were raised locally, and Larsen introduced reindeer for food and hunting. These have successfully reproduced on the island. The only women permitted on the station were the wives of the managers, and alcohol was prohibited. Captain Larsen became a dual British and Norwegian citizen and left the island a wealthy man in 1914.

Whaling came to an end at Grytviken in 1965. The development of offshore 'factory ships' some years before had made a land-based factory redundant.

In 1982 South Georgia was briefly front-page news when it became the first British territory to be invaded since the Second World War. On 19 March 1982 a Buenos Aires scrap metal merchant, Constantino Davidoff, landed on the island, ostensibly to junk the disused whaling station at Leith harbour. But almost immediately, Davidoff and his 50-strong team raised the Argentine flag – the Argentine government was later revealed to have been in regular contact with Davidoff. Three weeks later the island was recaptured by British forces, following the sinking of the Argentine submarine the *Santa Fe*. The invasion was a precursor to the Falklands conflict.

ABOVE Within weeks of landing, the 40-strong team of Norwegian whalers erected eight corrugated iron huts, workshops and sheds. The original accommodation block still stands.

OPPOSITE Chains rusting in the abandoned harbour: Grytviken's active life as a full-time whaling station lasted from 1904 to 1965.

FOLLOWING PAGE The old corrugated iron buildings, rusting in the bleak, beautiful landscape, are a sculptural legacy of a once thriving industrial town.

3 CHURCHES, CHAPELS AND MISSION HALLS

OPPOSITE AND BELOW Knolton Mission, Knolton Bryn, Overton, Wrexham (left) and St David's Church, Gaufron, Powys (below). By the end of the nineteenth century the archetypal corrugated iron chapel, unpretentious single storey timber-framed sheds, had become centres of worship for countless Christian denominations throughout Britain. Many of these modest buildings have survived.

Today churches, chapels and mission halls are among the most common and best loved of all corrugated iron buildings, but it was not until the last quarter of the nineteenth century that the trade in 'tin tabernacles' – one of the collective nicknames by which these buildings became known – really gathered pace.

A variety of denominations, including the Church of England, had by then explored the potential of corrugated iron as a construction material. From the beginning of the 1850s Anglican places of worship were built of corrugated iron in Britain's overseas colonies, at army camps and as temporary chapels while permanent premises were under construction. But these experiments were not universally popular. In the eyes of many within the religious and social establishment, corrugated iron was perceived as inappropriately ephemeral and aesthetically incongruous. For obvious reasons it also lacked any associations with tradition.

For dissenting congregations, which flourished in enormous numbers between the 1880s and the start of the First World War, these issues were less problematic. It might even be argued that the simple unadorned honesty of the material resonated positively with the manufacturing, engineering and mining backgrounds of the dissenting churches' predominantly working-class congregants.

During this period thousands of corrugated iron chapels were erected all over Britain. The many survivors stand as a physical legacy of the late-nineteenth-century religious revivals. Some are still in use, others are suffering the ignominy of dereliction, but all are decomposing with grace.

'Architectural Deceit'

From the outset of Queen Victoria's reign, religion occupied an unusually prominent position in the public consciousness. As a consequence, the design and construction of churches was taken extremely seriously and was the subject of impassioned debate.

Many commentators were offended by what they regarded as poor quality places of worship, some of which had been built in a hurry, using new materials and methods of construction, including corrugated iron. Much of the offence focused on the value system attached to the decoration of churches: ornamentation in the industrial age could be mass-produced, but mass-produced ornament lacked *gravitas* and integrity.

In 1841 in his *True Principles of Pointed or Christian Architecture* Augustus Welby Pugin, a deeply religious man and one of the most revered British architects of the nineteenth century, railed against dishonesty in architectural expression. He was particularly concerned about the quality of contemporary ecclesiastical architecture. In his view churches should be 'more vast and beautiful' than the homes of their creators, and he believed that too many early Victorians were failing in this duty, with richly decorated facades disguising cheap and 'false' rears. Pugin's views carried great weight throughout high Victorian society.

Another prominent force in the promotion of truthfulness in religious architecture, and a return to the purity of medieval values was the Cambridge Camden Society (from 1845 the Ecclesiological Society), which was formed by a group of Cambridge University undergraduates in 1839. During the 1840s, the society exerted a significant influence on the moral tone of society, and the Anglican Church in particular – members included bishops of the Church of England, Cambridge University deans and members of parliament.

John Ruskin, the noted social commentator and art critic, was also a prominent voice in the debate. In 1849 he dismissed any attempt to use 'modern materials' in the construction or decoration of churches, and campaigned vigorously for architects and craftsmen to oppose the use of corrugated iron and all other industrial impostors. 'The use of cast or machine-made ornament of any kind' is an 'Architectural Deceit', he wrote in *The Seven Lamps of Architecture*.

BELOW TOP The external walls and roof of Peter Thompson's temporary church in Kentish Town were covered with Croggon's Patent Asphalte Felt (sic), '[it] being the best non-conductor of hear and cold', according to *The Builder* (1844, vol.2, p.471).

BELOW BOTTOM Peter Thompson's 'Church for Jamaica', designed in 1844, was the first portable place of worship designed in Britain for export to the colonies made entirely of iron.

To many the views of Pugin, the Ecclesiological Society and Ruskin were object lessons in good sense, and help to explain the relatively late appearance of corrugated iron churches in Britain. However, they did not stop corrugated iron manufacturers looking for more responsive markets. Iron churches may not have been immediately palatable in Britain, but the rapidly growing communities of the colonies were another matter.

Iron churches for the colonies

The practice of making buildings in Britain and sending them out to trading settlements, missionary posts and other communities connected to the 'mother country' predates the invention of corrugated iron. During the late eighteenth century prefabricated timber churches were sent to Sierra Leone, South Africa and no doubt other locations of interest to the British. But these were rarities. It was more typical for churches to be built locally using whatever materials were available.

Between 1836 and 1851 at least nineteen churches were built in Melbourne with the idea that they would be replaced when skills and materials were available to build more enduring structures. But time, skills and resources militated against the construction of churches in the colonies and, inexorably, Britain's builders of prefabricated buildings started to turn their attentions to the creation of iron churches.

By the mid-1840s Peter Thompson, one of the early British manufacturers who designed portable buildings for the colonies, had already erected several temporary churches in and around London. But the church he supplied to Jamaica in 1844 represented the dawn of a new era in the supply of prefabricated churches; it was the first made almost entirely of iron. The *Illustrated News* of 28 September 1844 was enthusiastic:

> The pilaster supports are of cast iron, on which are fixed the frame roof of wrought iron, of an ingenious construction, combining great strength with simplicity of arrangement; the whole is covered in corrugated iron... The body of the church is 65 feet by 40; the chancel, 24 by 12; a robing room and vestry are attached... The cost of this iron church is £1,000.

IRON CHURCH, FOR JAMAICA.

ABOVE The second of Samuel Hemming's six churches for the Diocese of Melbourne on show at the manufacturer's Bristol works in the early 1850s.

GALVANISED IRON CHURCH FOR THE DIOCESE OF MELBOURNE,
SOUTH AUSTRALIA.

LEFT Another of Hemming's churches for Melbourne, mistakenly referred to by the *Illustrated London News* as a city in South Australia.

So when the Australian gold rush began in 1851 British manufacturers were ready to provide for the spiritual requirements of the huge numbers of diverse prospectors and hangers-on.

In 1853 the Diocese of Melbourne placed an order for six iron churches – enough for a combined congregation of 5,000 people – with Samuel Hemming of Bristol. As was common with manufacturers of prefabricated buildings at the time, the first of the churches was erected at Hemming's Iron Building Manufactory. It was both a promotional opportunity, and a chance to mark all the components so that they could be put together again in the right order on site. On Friday 13 May 1853 an inaugural service was officiated over by the archdeacon of Melbourne, then visiting England. Over 800 attended, no doubt attracted by the strange silver spectacle – paint was usually added once buildings had been erected in their intended location.

LEFT Church probably designed by Frederick Braby for South Africa, c.1883. The export of corrugated iron churches from Britain continued throughout the nineteenth century.

BELOW St Edmund's, Great Whale River, c. 2006. The church is in its fourth and almost certainly final location.

What they saw was a church for a congregation of 650–700, measuring 70 feet by 45 feet which adhered to principles of simplicity and utility in its design. It also included all the features common to traditional churches of that time, including a nave with side aisles, a baptistery and vestry and a prominent 75-foot tower. Hemming lined the interior in half-inch match boarding to be covered with canvas primed for papering. The roof and external walls were clad in galvanized corrugated iron. Packed for export, the church weighed 50 tonnes and cost £1,000. It had taken only five weeks to construct.

As he worked through the commission, Hemming's churches became more sophisticated. He used the low pitched corrugated iron roofs to create an Italianate look, and dressed the interiors in classical motifs. But in Britain his efforts found few supporters. One contemporary critic described the architecture as, 'hopelessly unpleasing... [it] suggests the factory or the warehouse'.

The export of corrugated iron churches to the colonies continued throughout the nineteenth century; wherever there were British territorial or trading interests, spiritual guidance followed.

In 1860, in the wake of the Fraser River gold rush, in a remote area of western North America that is now British Colombia, Hemming supplied a church with space for a congregation of around 600 to the small town of Victoria (see pages 100–103). A few years later he may also have supplied another smaller church to the Hudson's Bay Company, in Quebec. Remarkably, this small chapel – now in its fourth location – still stands, in the town of Great Whale River (also known as Kuujjuaraapik, Whapmagoostui and Poste-de-la-Baleine). It must be among the most isolated ever built; the town is nearly 825 miles from Montreal and over 165 miles from the nearest main road.

The story of St Edmund's of Great Whale River epitomizes the value of corrugated iron kit churches to frontier outposts. It began life in the early-1860s as a mission church in Fort George, Canada – missionary activity generally took place in locations not yet formally annexed by any colonial power. Fifteen years later, having served its purpose in that settlement, it was relocated to Little Whale River, the most northerly station of the Hudson's Bay Company. It was moved there at the request of

Edmund Peck, who had been engaged to proselytize among the local people, including Cree and Inuit. Prior to its arrival he held services in the igloo of a converted Inuit.

Unfortunately for Peck, the instructions for erection did not survive the move. He recorded in his diary:

> God has enabled me to erect the iron church... a nice, neat little building, measuring 40 feet long by 20 wide. I had plenty of puzzling work, as the ground plan could not be found; but with experiments, perseverance, and hard work, we managed finally to get everything in its place. The building was opened on Sunday, October 26. I preached in Eskimo, Indian, and English to my small flock.

Peck continued to use the little corrugated iron church until 1891, when it was dismantled, packed onto dogsleds and taken to the neighbouring settlement of Great Whale River, where it was erected for a third time – its current site is the second it has occupied in the town. It is now a museum, housing Inuit artefacts.

Early iron churches in Britain

Corrugated iron churches in Britain date from the mid-1850s. The first built in London – designed by Samuel Hemming – was St Paul's in the grounds of a vicarage in Kensington. Unlike Hemming's Australian churches, which had sparse external ornamentation and borrowed loosely from the forms of typical English village churches, St Paul's had a much less formal exterior, suggestive of a Swiss chalet. The roof of the nave projected beyond the gable wall, and the side aisles wrapped around the main volume of the church, creating an open porch across the whole width supported on ten slender columns with round-headed arches. This unusual feature broke down the external volume of the building, and created a generous area in which parishioners could gather.

St Paul's was remarkably original in its design, perhaps more so than later churches, which were either austerely utilitarian or attempted to imitate in iron the forms of traditional masonry buildings, with the addition of spires, pointed windows and other familiar features. So it comes as little surprise that it did not conform to expectations. In October 1855 *The Builder* (Vol. 13, p.507) remarked: 'It would not be difficult on a future occasion to give a more ecclesiastical character to such a structure externally.'

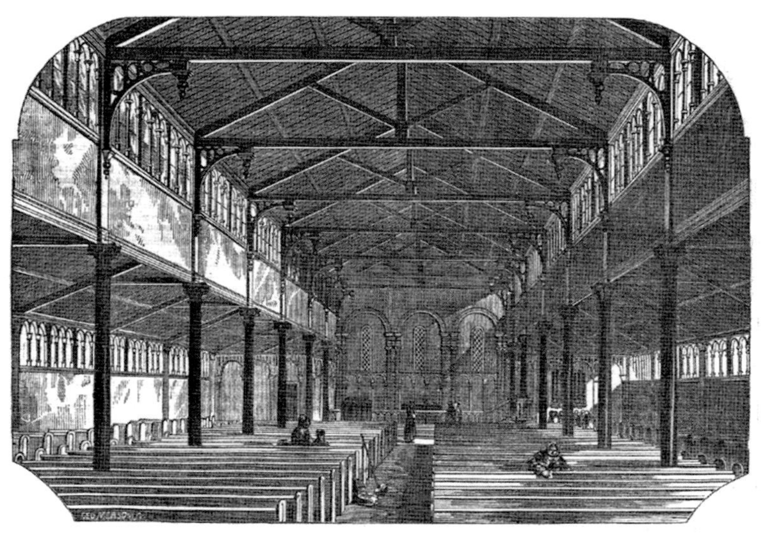

ST. PAUL'S TEMPORARY CHURCH, KENSINGTON.

ABOVE The exterior of St Paul's Kensington, the first corrugated iron church in London, was reminiscent of a Swiss chalet, an unusual departure for mid-nineteenth century British church architecture. Congregants of St Paul's each paid five shillings towards the cost of the structure. A permanent 'edifice of stone' was to be funded by any surplus.

OPPOSITE Dalswinton Mission Church, in Dumfriesshire, Scotland was built in 1881.

ABOVE St Augustine's Catholic Church, Draycott in Clay, Staffordshire, England.

BELOW The interior of St Nicholas, a well-preserved Church of England tin tabernacle in the Somerset village of Porlock Weir. It was typical for internal walls to be finished in match boarding. It was also common for manufacturers to provide all fixtures and fittings, even candles.

The British Army began building temporary corrugated iron churches in 1855. St Barbara's at Deepcut Garrison, built in 1901, is a rare survivor.

Also in 1855 three temporary churches were ordered for the army camp at Aldershot. At the time the Board of Ordnance was engaged in the expansion of army camps in response to the conflict in the Crimea (see Chapter 4). One of the three was an iron church, ordered from Hemming. A contemporary account describes it as, 'built entirely of... iron, with painted glass windows. It is particularly light and elegant and intended, we understand, for the Court especially'. Queen Victoria and her retinue worshipped there on her frequent visits to Aldershot to review the troops and to observe 'Sham Fights'. It was dismantled in 1926. The Army continued to use temporary iron churches for many years. St Barbara's, built in 1901 at Deepcut Garrison, still stands.

The late 1850s saw a rash of temporary iron churches in London. An iron church proposed in 1856 for Lambeth Walk, Brixton was welcomed by the Metropolitan Board of Works, who approved the building application on the grounds that it 'was designed in accordance with several former structures submitted by Mr Hemming'. In 1857 there was even some rare praise for iron churches. A commentator in the *Civil Engineer and Architect's Journal* wrote:

> The iron churches erected in London are found to answer every purpose for which they were designed. There are now five of them, in ... Kensington, Kentish Town, Newington Butts, St George's East and Holloway. The Holloway church cost £1,000, and is capable of housing 900 people. It is described as 'a most comfortable place of worship, well ventilated, warm in winter, cool in summer, and can be easily taken down when not needed in the district'. (Vol. XX, p.236)

During the autumn of 1858 a flurry of correspondence in the letters pages of *The Builder* exposed a broader range of views among its readers. An 'AR' initiated the exchange:

> Do you think that as iron has become a building material applicable for building purposes – to wit, churches – that it would be very desirable for some of the eminent firms in the iron trade to consult architects more generally as regards the design and general character of their structures? As it stands at present, I do not think there is a good iron church or school-house in existence.

LEFT AND RIGHT Although never built, the 'Iron Church' designed by Richard Carpenter and William Slater, and published in *Instrumenta Ecclesiastica* in 1856, represented the high water point of the Church of England's interest in corrugated iron.

Two weeks later a 'Cerca' responded:

> The iron abortions at Islington, Barnsbury, Bow and Kingsland ought certainly to be stopped, or a much greater improvement made; they remind one too much of a California mining district, where cheapness and mean utility is the paramount object to be obtained.

Hemming did not reply but Tupper and Co. (formerly Tupper & Carr) did. They claimed responsibility for building a church designed by the eminent architect Matthew Digby Wyatt for Rangoon, Burma, as well as others in Bournemouth, Islington and Dalston (see pages 91 and 98). They also launched a defence of their work: 'Everyone who has seen [the interiors of our churches] has commended them, and we are not afraid to compare them with any permanent church for good workmanship.'

From that point the exchange dissolved into a slanging match. 'Cerca' suggested that Tupper and Co. were guilty of failing to invest any effort in the design of the exterior of the Dalston church. This led Tupper and Co. to the conclusion that, 'Cerca wants a job… and he can have one if he will furnish us with a satisfactory design for an iron church, combining architectural beauty with cheapness of price.'

The squabble gives some sense of the extent of concern about the use of corrugated iron in church buildings. Nevertheless, by the mid-1850s, it was clear that 'tin tabs' were in Britain to stay. So the Church of England made one final effort to embrace the material.

In 1855 Richard Carpenter, an architect and member of the influential Ecclesiological Society, began work on the design of an iron church. When he died a year later, his pupil William Slater continued the project. The outcome was published in 1856 in *Instrumenta Ecclesiastica*, the second volume of a compendium edited by the Ecclesiological Society, of approved designs for everything needed for an Anglican church, ranging from chalices to whole buildings prepared by leading architects. The society was clearly not much impressed with what had gone before:

> It may safely be assumed that the iron churches, of which several have been sent to the colonies, or erected as temporary churches at home, have not… [been designed] in accordance with the qualities and conditions of the material. Their framework is of wood covered externally with corrugated iron; the pillars are wooden posts and the roofs both of nave and aisle are wooden in their construction. What is such a building but a wooden structure encased in metallic plates? The iron structures so familiar to our eyes in railway sheds are altogether unecclesiastical in character and associations. They fall within the province of engineering rather than of architecture.

The description published with Carpenter and Slater's design stated that:

> The external walls are a frame-work of cast iron, so arranged as to have the interstices faced internally and externally with corrugated plates, and packed between these plates with felt and sand. The arches (lateral and transverse), the framework of the roofs, and the girders of the aisles are formed of iron castings riveted together.

The design marked the zenith of the Church of England's interest in corrugated iron, but it was never built.

Common Interments from 7/6.
Private Graves from 15/6.

RIDGEWAY P

K CEMETERY.
THOMAS OATES, Registrar.
45, Park Street St or
THE CEMETERY. (Bristol.)

PREVIOUS PAGE An advertisement for Ridgeway Park Cemetery, Bristol, *c.*1888. The corrugated iron chapel, on the right, was supplied by local manufacturer Messrs J. Lysaght Limited, a company that later flourished in Australia. The developers planned to replace it with a permanent chapel as soon as funds were available, but it survived until 1951.

LEFT From the 1880s, almost every manufacturer involved in the construction of iron building products included churches within their repertoire. David Rowell & Co. of Westminster was no exception.

OPPOSITE St Mark's temporary iron church in Dalston, East London, designed by Messrs Tupper and Co, was severely damaged by a tornado in 1865.

The age of tin tabernacles: 1860s–1920s

It was not until the second half of the nineteenth century that the production of corrugated iron church buildings really took off. There were several reasons, including the demographic shift caused by the Industrial Revolution, which created new communities very quickly, the bulk of which had no spiritual meeting points; the growth of non-conformist sects, many of them supported by the new urbanized and increasingly literate working classes; and the Church of England's on-going building programme, which created a demand for temporary premises while construction of permanent stone structures was underway.

By the 1860s manufacturers of corrugated iron buildings were well equipped to meet the growing demand. Some were large, established companies catering for countrywide, even international, markets. In London these included, Samuel Hemming, Tupper & Co. and Walker's Corrugated Ironworks (from the mid-1860s the company founded by Richard Walker in 1829 was run by W. H. Griffith, and later M. H. Davies). Liverpool was the other major centre of production. Local manufacturers included: Isaac Dixon and Francis Morton & Co. Others were smaller operations, catering principally for local markets.

From the number of advertisements in the technical press and trade directories, it is clear that from the early 1880s the number and scale of manufacturers spiralled. By then just about every company in or on the fringe of the corrugated iron industry was marketing itself as a purveyor of church buildings, schoolrooms or mission halls. Some of the most successful companies to evolve from this frenzy of activity were Boulton and Paul of Norwich, Frederick Braby & Co. Ltd, Liverpool, and William Cooper Ltd and David Rowell & Co., both of London.

Temporary churches were paid for in a variety of ways. For congregations with limited budgets, it was possible to rent premises directly from manufacturers. In 1856 Samuel Hemming promoted, 'iron churches similar to those erected by the subscriber in England and the colonies for sale and for hire' in *The Times* of 28 April. An alternative approach was to buy a church that had served its purpose, such as the 'temporary iron church on the site of Christ's Church on Penton Street, [Islington],' which was sold at auction in September 1860 (*The Times*, 3 September 1860). Another auction was held for St Mark's temporary church in East London, which was severely damaged by a tornado on 22 August 1865.

> The well timbered and corrugated iron church, erected by Messrs Tupper and Co, at a considerable outlay... [will be] sold on February 21, at 12 o'clock, in one lot. [It is] to be paid for and cleared away within 14 days. (*The Times*, 17 February 1866)

The timing of the tornado could have been worse. By 1865 the temporary iron church had been standing for five years, and its permanent replacement was nearing completion. Today, St Mark's Dalston, a bulky edifice with Gothic trappings, is known locally as the Cathedral of the East End'.

For new churches it was common to base prices on the seating capacity. In 1867 Francis Morton & Co. erected St Marks at Claughton, Birkenhead at a cost of £4 per sitting or £2,000 for 500 people. In its advertising literature the company made much of its design – its 'Church Building Department having recently been placed under the management of an experienced architect'. Morton even suggested that its, 'Churches, Chapels, School Houses or Lecture Rooms [could fulfil] either temporary or permanent uses'.

Another function of temporary churches, aside from keeping congregations together while permanent premises were under construction, was as a means of generating revenue. Some congregations, particularly in wealthy areas, could be relied upon to make significant contributions towards the costs of the permanent replacement. In a highly stratified society, philanthropists also took it upon themselves to play a role in maintaining the standards of the masses. In 1863,

in answer to the appeal of the Bishop of London to owners of property in London to contribute towards the spiritual destitution of the metropolis a novel response [was] made by Mr. G. Cubitt, M.P. A committee of gentlemen formed to enlarge the temporary iron church of St Clement's, Barnsbury [London] applied for his assistance when he most munificently offered to endow the district with £150 a year... entirely at his own expense. (*The Times*, 12 December 1863)

Dissenting sects

Of course, it is with dissenting sects that tin tabernacles are primarily associated. The vast majority of the chapels, mission halls and churches that have survived were for nonconformist congregations. The extent of contemporary demand is emphasized by the findings of the 1851 Religious Census, which revealed that there were already more nonconformists than Church of England worshippers. It also showed that many of them had nowhere to worship. For instance, in 1851 Hulme, an industrialized district of Manchester whose population increased fifty-fold during the first half of the nineteenth century, had a population of 70,000, but pews for only 7,500. The pattern was repeated all over Britain.

Initially many of these groups met in the front rooms of private houses, or the back rooms of pubs. Many of the Temperance meetings took place in tents erected in the field of a sympathetic farmer. But gradually the groups of worshippers managed to collect the funds to buy one of the new corrugated iron chapels available in kit form from manufacturers.

From the 1870s Portsmouth's rapidly expanding dock district became home to a number of corrugated iron churches, chapels and mission halls. The location and lifespan of the buildings reveals much about the pace at which the nonconformist churches moved. In 1892 a second corrugated iron church was built on Albert Road, to service the area's strong Methodist community – the first had been built in 1864. The new church was erected by a local builder, Mr Harbour, at a cost of £239. But it did not last long. In 1900 it was sold at auction for £50 and relocated to Eastney Road, where it was put to use as a Primitive Methodist Chapel.

The trade in recycled iron churches was under constant scrutiny from agents trying to promote new models. St Mary's in the Kingston district of Portsmouth, initially built in the twelfth century, was demolished in 1887. While its replacement was under construction, a temporary iron church was built on the edge of the site. In 1889, the new stone church complete, the redundant tin tab was offered to the Victoria Road United Reformed Church for £250, plus the cost of dismantling it. Negotiations took place through an agent. But while the United Reformed Church was considering its options, Messrs Compton and Faukes Company, a manufacturer of iron buildings, offered it a new iron church at a comparable price with a £25 donation to sweeten the deal. There is no record of the agent's share of the profits.

KNOWLE

PREVIOUS PAGE Knowle Mission Church in Shropshire was the Clee Hill Granite Company's response to fears about the drinking habits of its employees.

ABOVE Victoria Road Congregational Church, Portsmouth, home to the United Reformed Church of Victoria Road. The structure was designed by local manufacturer Messrs Compton and Faukes Company.

LEFT The second Methodist 'tin chapel' on Albert Road, Portsmouth, lasted for eight years before it was acquired by a congregation of Primitive Methodists.

A common reason for the construction of nonconformist churches was the bawdy behaviour of employees in areas with no recent history of industry. Bailbrook Mission Church in Bath was built to bring Christianity to the employees of Robertson's fruit orchards, who had a reputation for gambling and drinking. The church, supplied by William Cooper Ltd at a cost of £200, was consecrated on 4 July 1892. Today, in an advanced state of decay, Bailbrook is a private home.

Another church bought by a company concerned for the spiritual wellbeing of its employees still stands on Clee Hill in Shropshire. Quarries at Titterstone Clee Hill have supplied stone, coal and lime to the surrounding areas for centuries. In the 1860s companies were formed to exploit these resources. The following years saw a rapid influx of men from the Black Country, Wales, Nottinghamshire, Yorkshire and Scotland. Knowle, a hamlet on

ABOVE Bailbrook Mission Church in Bath was erected to help tame the bawdy behaviour of employees at local orchards.

FOLLOWING PAGE Colonization and mineral exploitation have left a diverse legacy of corrugated iron religious buildings. In the 1870s, while prospecting for gold in Queensland, James Venture Mulligan stumbled across extensive tin deposits. A large Chinese community was among those that took advantage of the find. They also built the Hou Wang Temple, a timber-framed structure clad in corrugated iron, which was restored to its original glory in 2002.

the side of the hill grew particularly quickly. Responding to fears about the amount of alcohol consumed by the miners, and the lack of spiritual guidance available in the area, the Clee Hill Granite Company donated a small iron church to the hamlet. The Knowle Mission Church was built by the miners themselves and completed in the 1880s. It is still in occasional use.

The genesis of the Labour Movement

If the 1890s represented the high-water mark for tin tabernacles, the First World War marked the beginning of their decline. From the early-1920s, the popularity of religion began to wane; the social upheavals of the Industrial Revolution also began to settle down.

Once again the manufacturers of corrugated iron structures proved themselves to be fleet-footed in response to prevailing trends; farming and light industry became particularly vibrant markets during the inter-war years. But the legacy of the religious revivals lives on. Countless corrugated iron meetings halls, chapels and churches still stand in towns and villages across Britain, and other corners of the globe. The legacy of the phenomenon also survives in the less obvious form of the British Labour Party.

Keir Hardie, Britain's first socialist member of parliament, was a lay preacher for the Evangelical Union Church. He was also involved with the Temperance Society, and from the late-1870s was an activist in the burgeoning trades union movement. Throughout his career his activism was informed by his religious beliefs. For many years before his election to parliament, he travelled across Britain promoting the rights of working people from corrugated iron churches, mission and village halls. He finally became an MP in 1892, winning West Ham as an independent socialist candidate. The following year Hardie and colleagues formed the Independent Labour Party and, in 1900, brought many union-based and other socialist groups together under the banner of the Labour Representation Committee, renamed, in 1906, the Labour Party.

The Independent Labour Party was incubated in tin tabernacles. It is no exaggeration to argue that the present-day Labour Party would not exist without the invention of corrugated iron.

SIGHT OF ETERNAL LIFE CHURCH

Shrubland Road, Hackney, London

OPPOSITE PAGE The former Congregational Church in Hackney, East London, built in 1858, may be the world's oldest surviving corrugated iron church.

Corrugated iron churches were a common feature of mid-nineteenth-century Hackney, a period that saw the borough's transformation from middle-class suburb to poor inner-city area. Examples included St Marks on Ridley Road, St Mary of Eton in Hackney Wick, St Augustine's on Dalston Lane and St Matthew's in Upper Clapton.

Over the years the majority were either replaced with permanent structures, or fell into dereliction pending removal, a common outcome for nonconformist groups. Only one of the iron churches has survived, the former Congregational Church on Shrubland Road, known today as the Sight of Eternal Life Church. Built in 1858, it may be the oldest surviving corrugated iron church in the world.

The Shrubland Road church was founded by Presbyterians in the Revd Thomas Whyte's congregation. The Reverend negotiated an 80-year lease for the site at an annual rent of £15 from the Rhodes estates, the local landowner, in July 1858.

Nonconformism in Hackney has a long history. Several dissenting organizations were established there. The Presbyterians go back the furthest (1636), but the most unusual is more recent. The Agapemone, or 'Abode of Love', was established in Clapton by Henry James Prince, a self-styled second messiah, who promised his adherents eternal salvation of the body and soul. Prince, a colourful character, survived several sex scandals before he died – much to the surprise of his congregants – in 1902.

An application to dig the foundations and lay a drain for the Shrubland Road church was made by W. Browne of Messrs Tupper and Co. (formerly Tupper and Carr) of Moorgate Street, a short distance south, on 24 August 1858. Tupper and Co., one of the pioneers in the production of iron buildings, promoted its wares at the Great Exhibition of 1851, alongside Edward Bellhouse, Morewood and Rogers, Charles D. Young and Co. and others. In its early years the company was well known for the manufacture of components of buildings, but it did not take long for it to diversify into whole buildings. From the 1860s 'Tupper & Co. Iron Church and House Builders' were among the leaders in the field of iron churches, and they had been quick off the mark.

The first recorded church in Britain was built in 1855, but by 1858 Tupper and Co. had already completed a 'Scottish Church' in Bournemouth and a large tabernacle in Islington, London. The company also supplied St Marks on nearby Ridley Road (see page 91). It was also one of the first manufacturers to employ the services of an architect, George Adam Burn, although it is not known whether anything came of this partnership.

It seems certain that he was not involved in the design of the Shrubland Road church. The only features that give the 37-feet by 72-feet oblong structure any sense of ecclesiastical character are its 48-feet-high belfry and pointed Gothic-style window frames. In other respects the building is a large corrugated iron clad shed with a pitched roof. The interior, however, is more refined. The timber panelled entrance features chamfered muntins. A reading platform with twisted balustrades, curved pulpit and large colourful organ define the light-filled congregation space.

Nevertheless some people were impressed by its outward appearance. Mr J. Cox, an historian who, over twenty years, compiled a handwritten 508-page account of 'notes relating to Shoreditch', stated: 'As an example of celerity it may be mentioned that the church was erected in the short space of ten weeks... It will accommodate 500 persons, cost £1,200-1,300 and is a very handsome structure.'

The original congregation does not appear to have prospered. By 1871 the worshippers were describing themselves as Congregationalists rather than Presbyterians. Things took a turn for the better in 1878, when the Revd Thomas Udall began his ministry. The next thirty-one years saw the most successful period in the church's history.

The Revd Udall inherited a congregation of around 50–60. By the time of his death in 1909, the church had a Sunday school with over 300 members, a Band of Hope (temperance society for working-class children) with a membership of 250, a flourishing social guild and a men's sick and provident society with 530 members. But despite its popularity, the Revd Udall was unable to raise the funds to build a permanent church.

On 1 December 1909, to mark the (slightly belated) fiftieth anniversary of the church, the *Hackney and Kingsland Gazette* recorded:

> A serious difficulty now confronts the earnest band of workers [congregation]. The church requires a new roof and heating apparatus, and sanitary and general repair, involving a total outlay of £400. Such a sum cannot be raised in the poor neighbourhood in which the church is situated.

It is not uncommon for nonconformist sects to struggle to raise funds; it is a problem that spells the end for many church buildings. Fortunately the problems were not terminal at Shrubland Road. The church also had a relatively large congregation to help keep the building alive, which may go some way to explaining how the building survived for so long after the Revd Udall's death. Even so, the numbers did eventually tail off.

By the late 1960s the end was nigh, and in 1971 the remaining faithful merged with Trinity Chapel in Lauriston Road, in South Hackney. That same year the building became home to the Sight of Eternal Life Church, an evangelical sect. It remains in their ownership today.

In 1975 the church was listed Grade II, ensuring that its future is secure.

ST JOHN'S CHURCH

Victoria, British Columbia, Canada

Corrugated iron churches came relatively late to British Columbia on Canada's western seaboard, but when the first arrived no expense was spared. Sadly, St John's was not long for this world.

In the 1850s the small town of Victoria had a population of trappers, fur traders, native Indians and the British Navy. It was also home to the Revd Edward Cridge, chaplain of the Hudson's Bay Company, who had a congregation of about 400.

Victoria was a strategic outpost of the British Empire – the navy's presence helped consolidate British claims to the territory. Revd Cridge had been employed to lend spiritual sustenance to the small European population. He also embarked on missionary work with the local Indian population. Until 1858 life ticked along smoothly enough, but then the glint of gold briefly lit up the western Canadian mountains.

The Fraser River gold rush attracted a massive influx to Victoria, mainly from San Francisco. Within the space of a few months the town's population leapt from 500 to almost 8,000. The Revd Cridge was no longer in a position to service the population's spiritual requirements, so he wrote to the Colonial Church and School Society – whose purpose was to support Anglican churchmen, missionaries and schoolteachers overseas – and asked for help. Some money to fund the expansion of his church and build a rectory was probably the extent of his expectations. He could not possibly have predicted the response. His letter was passed to Angela Burdett-Coutts, then the wealthiest woman in Britain. Without hesitation she agreed to endow the church with £25,000 for a bishopric and two archdeacons in British Columbia and pay for the erection of a new church.

Burdett-Coutts inherited her grandfather's £2 million banking fortune at the age of twenty-three. She spent the rest of her life giving it away. Her projects were numerous and included the Ragged School's Union, the Temperance Society and the National Society for the Prevention of Cruelty to Children. She was also active in building Anglican churches and in 1847 endowed bishoprics in Australia and South Africa.

It was through her work for the Australian colonists that she met Caroline Chisholm, an army wife and an active voice for British colonists there. And it was through her that she was introduced to Samuel Hemming, the successful manufacturer of prefabricated corrugated iron cottages, portable houses, warehouses, shops, bazaars, hotels and churches for the colonies. He had already built corrugated iron churches for Melbourne (see page 81), for Aldershot barracks and perhaps for South Africa.

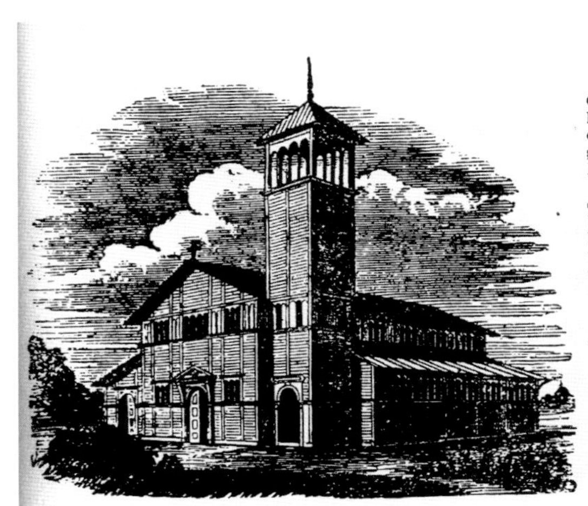

Samuel Hemming advertisement for iron churches, published in *Building News*, 22 May 1857.

The church for Victoria, British Columbia followed Hemming's usual style. A wide shouldered monolith built on his time-honoured pattern, it was to house a congregation of 600 people and at 60 feet long by 45 feet wide was slightly smaller than the first of the churches he had sent to Melbourne. The structure was framed in Baltic fir and clad entirely in galvanized corrugated iron. It had a nave and side aisles and all the other features that would have been expected of a fully functioning church in the mid-nineteenth century.

It was built through the summer of 1859 at the Bow Works and was inspected there in the autumn by the new Bishop of British Columbia, the Right Revd George Hill, while he was in the process of raising further working funds (a total of £11,500) prior to his departure for the outlying colony. A few minor alterations were made to the structure that brought the total cost of the church to just over £1,700 – shipping would cost another £900. Burdett-Coutts had even paid for the inclusion of pews, other furniture and a small organ. The latter, incidentally, had been earmarked for a new church in Nice built for the wealthy British wintering on the French Riviera, but Burdett-Coutts diverted it to British Columbia – perhaps a more deserving cause.

Hill and the church left England at the end of 1859. Hill arrived in January 1860 after a 50-day journey via Cape Horn and the church, on the barque-rigged clipper, *The Athelstan*, two months later.

St John's Church and rectory under construction in 1860. Samuel Hemming provided everything except the tower, which was built locally

Accompanying the building components were two men supplied by Hemming to direct and assist in its erection on the island; one of them was his son, John. Between them, and with the assistance of local builder John Cochrane, the church was built at the north-east corner of Douglas and Fisgard Streets, overlooking the growing town.

The cornerstone was laid during a ceremony on 13 April 1860 and the church was consecrated exactly five months later – the tower took a little longer to complete. Impressions were mixed. One report related that a passer-by assumed that it was a new brewery, whereas a gushing local reporter acclaimed it as the finest church north of San Francisco.

That winter problems became apparent in the church's construction. The unfinished tower allowed the wind to enter the roof space and, in addition to making the building cold and draughty, it lifted some of the roofing sheets, loosening them and allowing rain to drip on the congregation below. The biggest problems, however, were financial.

The Right Revd Hill's ambition was to make the church financially independent and this was to remain a problem throughout its history. Nevertheless, the stalwart members of the Ladies Guild invariably came up trumps. They provided decoration, furniture, new fixtures and fittings, they had the church lit with gas and had even raised the money to build a rectory house adjacent.

In 1868 a new vicar, Mr Jenns, made a few alterations. He found the interior of the church too sombre for his taste and brightened it up with the addition of stencilling on the stark interior timber boarding. Meanwhile the church's fortunes rose and fell with those of the colony. In 1872, following a period of prosperity, the church was operating in the black for the first time. This did not last but, sharing in the affluence brought about by a new rail link, the accounts were in the black again during the late 1880s. This was timely, as problems with the structure of the church were becoming apparent. The wooden sills upon which the foundations were laid had rotted, which left a small lake underneath the church whenever it rained – an expensive drainage ditch would eventually be required.

One aspect of the church that was never fixed, however, was its acoustics. One observer wrote:

> If a storm came up during the service, the wind rattled the iron roof like thunder... At the peak of the storm there would be moans and shrieking noises and only the organ could compete with the sound... The minister didn't have a chance.

As the nineteenth century came to a close, an ever-increasing repair bill led to discussions about the replacement of the iron church with a permanent masonry one. The freehold plot was finally sold in 1901. The final service was held on 15 December 1912, at which Mr Jenns had the last word: 'Leave the old landmark to the contractors to dispose of as they deem best,' he said. The church was pulled down a few weeks later.

ABOVE St John's decorated for Christmas 1889.

OPPOSITE The church towards the end of its life; it was demolished in 1913.

THE ITALIAN CHAPEL

Lambholm, Orkney, Scotland

BELOW The Italian POWs used concrete, scrap and salvaged materials to create the fittings and finishes, including the elaborate altar.

The Italian Chapel on Lambholm in Orkney, northern Scotland, was converted from two Nissen huts during the latter years of the Second World War. Beautifully and endearingly decorated by Italian prisoners of war from salvaged materials, it has been preserved as a reminder of Camp 60.

During the War, Scapa Flow, a huge natural harbour off the south coast of Mainland Orkney, housed the British Home Fleet. Heavily protected, it was considered safe from attack. But late in 1939 a German U-Boat managed to slip through the defences and sink the battleship *Royal Oak*, killing over 800 of her crew. At the War Office in London the Admiralty was shocked at the news and immediately commissioned a plan that would permanently block off the eastern approaches to Scapa Flow. Together with the construction company Balfour Beatty, it was decided to create over a mile of concrete barricade linking several of the islands with Mainland Orkney. This was to become known as the Churchill Barrier.

In May 1940, soon after work had begun, around 1,300 Italian prisoners of war, captured during fighting in North Africa, were shipped to the islands to assist in the barricade's construction. A potential breach of the Geneva Convention that did not allow POWs to take part in military projects was avoided by describing the building works as a civilian operation to create causeways linking the southern islands to Mainland Orkney.

Camp 60, on Lambholm, housed around 550 of the prisoners in thirteen Nissen huts. They arrived there from the island's main camp on Burray in January 1942, and almost immediately they began to

The chapel is composed of two Nissen huts placed end to end.

domesticate their surroundings. They built gardens and paths using leftover concrete. They constructed a theatre and a recreation hut, complete with a concrete snooker table, and – the only other evidence of the prisoners' residence on the island – a statue of St George slaying the dragon created out of concrete, barbed wire and wire netting.

But the greatest undertaking at Camp 60 was a chapel. Towards the end of 1943 two Nissen huts were joined end to end and work started on the construction of a church. Managed by Domenico Chiocchetti and aided by electricians Primavera and Micheloni, blacksmith Palumbo and artist Bruttapasta – whose skills in cement sculpture were renowned in Italy – the creation took almost a year and a half to complete.

Work began on the chancel first, followed by the altar, altar rail and stoop. All were constructed from concrete and scrap materials. Behind the altar Chiocchetti created his masterpiece, a painted window portraying the Madonna and Child, his symbol of the peace that he wished for. The metalworkers made candelabras from scrap metal and the rood screen took four months to create from steel reinforcing rods.

The chapel was by now in use but the Nissen huts' conversion did not stop there. The rest of the walls were panelled with plasterboard and a *trompe l'oeil* scene of Italian churches was painted. More candlesticks were made from brass salvaged from wrecked ships. These also provided the wood for the tabernacle. Lanterns were crafted from corned beef cans and the wire to hold them was stripped from ships' steel hawsers.

Attention then moved to the exterior. Now it was the turn of the cement artist, Bruttapasta. He fashioned a façade with a colonnaded porch; he also cast gothic pinnacles in clay. Sergeant Pennisi, who had been seconded from another POW camp by Chiocchetti initially to help design the front of the church, created the head of Christ. Coloured window lights were added and, other than a suitable bell that came later, the transformation of the Nissen huts was complete.

The prisoners used the chapel for only a short time before they were repatriated early in 1945, although Chiocchetti stayed on for a little to complete the font.

The islanders, recognizing the symbolic importance of the church, formed a preservation committee in 1958 and in 1960 they held a rededication ceremony with Chiocchetti in attendance. Further restoration work has been carried out since to the decorative work, but the original fabric, the corrugated iron cladding on its steel frame, has never needed any repairs.

4 WARFARE: HUTS, HANGARS AND HOSPITALS

Warfare spawned what is arguably the most famous of all corrugated iron buildings: the Nissen hut (see pages 118–23). From September 1916 until the end of the First World War more than 100,000 of Peter Nissen's barrack huts were erected close to the front in France and Belgium, providing accommodation for nearly 2.5 million Allied soldiers. Over the next sixty years, countless more were built around the world. During the Second World War the Nissen also inspired an American imitator: the Quonset hut (see pages 124-29).

The countless variations on Peter Nissen's original principle, of an arch made of curved and corrugated steel sheets attached to semicircular T-section steel ribs, served a variety of purposes in times of both war and peace. Many are still in use today, converted into stables, holiday homes, garages, workshops, even churches. But the story of corrugated iron and the military encompasses a great deal more than portable field accommodation, and dates back much further than 1916.

After the Royal Navy's pioneering use of corrugated iron in the roofs of slips during the 1830s and 1840s (see Chapter 1), more than six decades would pass before the British military adapted the material on a mass scale. Britain did not have a standing army stationed at home, so had no pressing need for troop accommodation. The second half of the nineteenth century was also a relatively peaceful period, with few military commitments apart from suppressing insurgencies in Africa, India and New Zealand, and projecting military force more widely in places where Britain had no territorial claims but trading interests, such as China. Even so, the time it took for the relationship between corrugated metal and the military to flourish does seem strange. To our eyes, its portability, versatility and relatively low cost seem ideally suited to a wide range of military applications.

OPPOSITE Members of the Royal Australian Air Force constructing an 'Igloo Hangar' in September 1943 on Goodenough Island, Papua New Guinea. By the end of the Second World War corrugated iron was an established feature of military life throughout the world.

BELOW A British Expeditionary Force hospital, including Peter Nissen's hospital hut, set among sand dunes in France (1917). Over 10,000 hospital huts providing 240,000 bed spaces were erected behind the lines during the last two years of the First World War.

One of the earliest known uses of corrugated iron in a military context was in the roofs of the auxiliary buildings for Renkioi Hospital, Turkey (1855), designed by Isambard Kingdom Brunel.

The Crimean War

By the early 1850s British manufacturers had produced prefabricated portable corrugated iron buildings for a wide variety of applications and climates. It was an advanced and technically proficient industry perfectly capable of turning its hand to military needs, whether in the production of storage facilities, barracks or field hospitals. So why, in March 1854 when the British joined Ottoman and French forces fighting the Russians in the Crimea, did their army not make greater use of corrugated iron? Other nations had recognized the merits of the material, including the American army, which in 1849 commissioned Liverpool engineer John Grantham to supply it with a barrel-vaulted iron barracks and a large corrugated iron warehousing system.

One reason was that the British did not expect the campaign to be protracted, so preparations for support services were not as thorough as they might have been. Other reasons included competition from timber and canvas, both well established features of life in the field, and the conservatism of the British military. However, the material was not completely overlooked during the Crimean War. One of the earliest known applications of corrugated iron in a military context was in the roofs of the auxiliary buildings at the hospital at Renkioi, designed by Isambard Kingdom Brunel.

At the beginning of 1855 the British army decided to build new hospitals in Turkey, ostensibly to relieve pressure on the hospitals in Scutari, a district of Istanbul (now Üsküdar) on the Asian shore of the Bosphorus. The decision was also driven by the inadequacy of existing hospital provision, a fact highlighted by Florence Nightingale, a pioneer of modern nursing, sanitation and hospital planning, who had been commissioned to oversee the introduction of female nurses to the Turkish hospitals.

Improvised local structures and prefabricated timber buildings went some way to addressing the shortfall, but not far enough. So the services of Britain's premier engineer were called upon. By this late stage in his career (he died in 1859), corrugated iron was well known to Brunel. He was fresh from making extensive and innovative use of it at Paddington Station (see page 24), which opened in 1854, and he was still on the board of the Galvanised Iron Company.

At Renkioi, Brunel's challenge was to design a prefabricated hospital that was cheap, portable, adaptable to all types of terrain, capable of accommodating as many beds as were required and built to the highest standards of sanitation. His solution was a hospital composed of independent wards and auxiliary buildings linked by walkways.

The wards were made of timber and covered in extremely thin and highly polished tin to reflect the sun. The auxiliary buildings were made of iron, including corrugated sheets, as a precaution against fire. According to an 1857 report they included: 'An iron kitchen, slightly detached from the wooden buildings fitted with every contrivance capable of cooking for 500-1,000 patients [and]... a similar building of iron ... fitted up with all the machinery needed for washing and drying.' Surviving images show structures with curved barrel-shaped roofs with over-hanging eaves.

In 1858 a report by military engineers considered whether prefabricated hutting – whether of timber or corrugated iron – would have been an asset during the Crimean conflict. The answer was 'no', largely because too many things could go wrong, including parts going missing and huts being poorly assembled. A similar report in 1872, following the Franco-Prussian War, came to the same conclusion. It was clear that the portability and familiarity of tents would take some beating.

War as a business opportunity

While corrugated iron may not have been used extensively in the field during the Crimean conflict, the home market was another matter. The construction of camps for Britain's mobilization was underway before troops even left for Turkey, and corrugated iron featured prominently within them in a variety of applications, including prisons, barracks, churches and buildings for recreation. Most of the buildings were designed by the same manufacturers

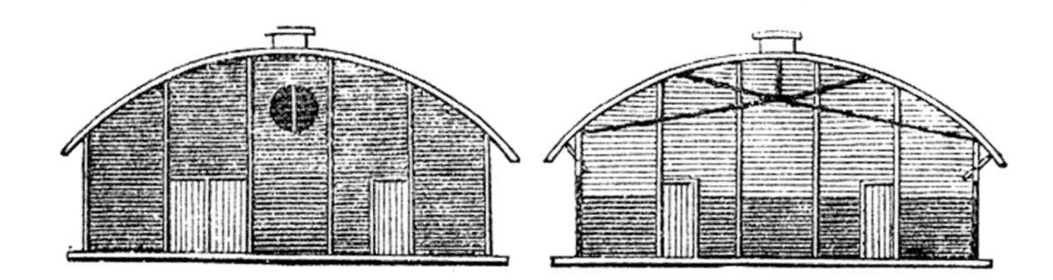

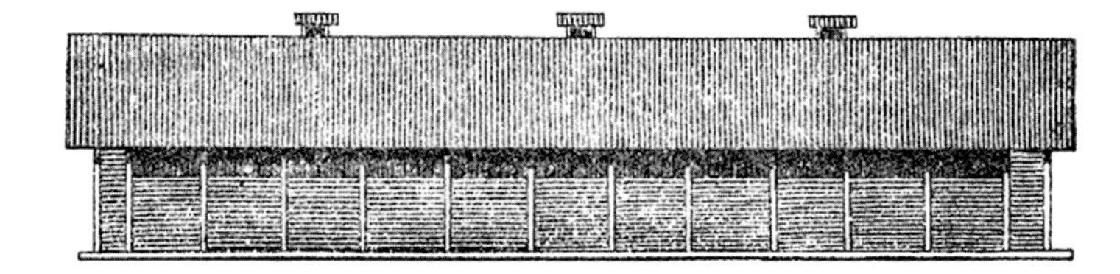

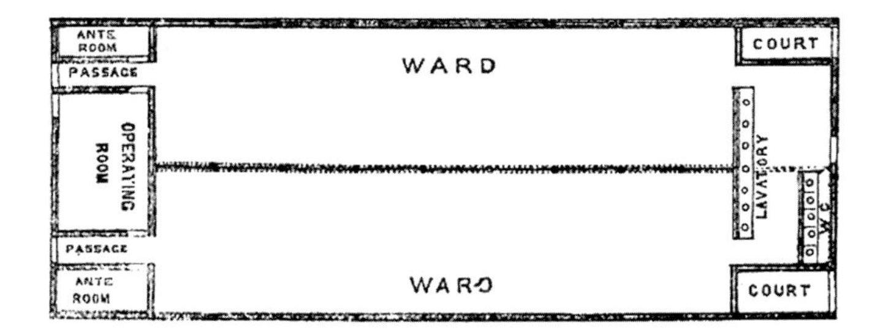

LEFT The corrugated iron Military Hospital promoted in C. D. Young & Co's catalogue of 1856 may have been an attempt to capitalize on interest in field hospitals generated by the Crimean War.

BELOW TOP The corrugated iron clubhouse at Aldershot army camp *c.*1855 was the height of luxury compared with the canvas field accommodation provided to personnel then on the front line in the Crimea.

BELOW BOTTOM Corrugated iron manufacturers were quick to develop products for conflict situations. Among Samuel Hemming's designs was a partially submerged bunker for Chalons, a training camp built by the French 'Second Empire' every summer from 1857 to 1870 in Champagne.

then supplying portable buildings to colonists, gold prospectors and anyone else in need of temporary or portable accommodation. To these entrepreneurial firms conflict represented a business opportunity.

On 17 February 1855 John Walker placed an advertisement in *The Builder*. His company was offering, 'portable galvanised or painted corrugated iron dwellings, barracks, hospitals, stables etc,' and he claimed that his clients included the governments of Britain and France. The previous year Walker had tendered for the supply of iron barracks for the army's camp at Aldershot. By 1857 he had supplied 41 buildings of three different types for a total sum of £5,112.

Charles D. Young & Co. also supplied corrugated iron 'Barracks, Cooking-Houses and Straw Stores' to the camps at Aldershot and Colchester. According to the company's promotional literature, they were all received to, 'the entire satisfaction of the Government Engineers' (Charles D. Young & Co., *Illustrations of Iron Structures for Home and Abroad*, 1855). There is, however, no record of the government having commissioned Young's military hospital, which featured prominently in the company's 1856 catalogue – perhaps the design was developed in the hope of building on the post-Crimean interest in iron military hospitals.

Also at Aldershot, and possibly other army camps, were some of the earliest iron churches built in Britain (see Chapter 3), at least one of which was designed by Samuel Hemming, who also manufactured the Royal Aldershot Clubhouse, a large corrugated iron clad structure composed of three gable-roofed blocks 82 feet wide and 132 feet deep. Construction of the clubhouse commenced on 28 July 1855 and was complete a month later. It contained rooms for reading, taking coffee, playing billiards and smoking; absolute luxury compared with the tents provided to those at the siege of Sebastopol.

Of all those involved in making corrugated iron buildings, Samuel Hemming alone seems to have succeeded in rivalling the major timber hut manufacturers. Hemming's many government contracts, for an impressive variety of buildings, amounted to just

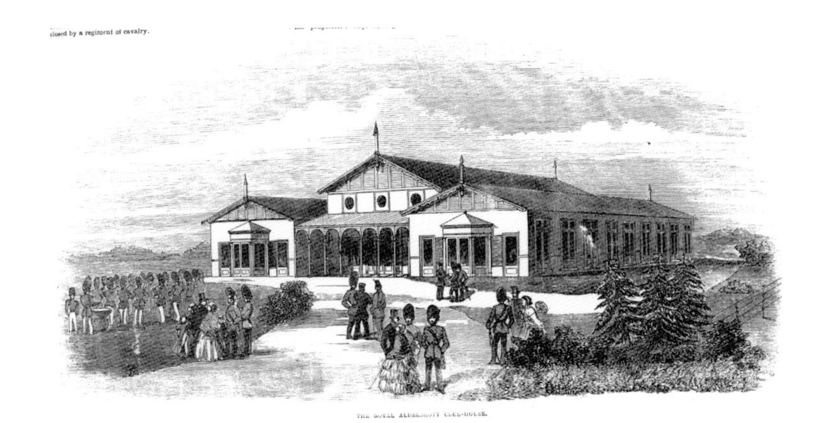

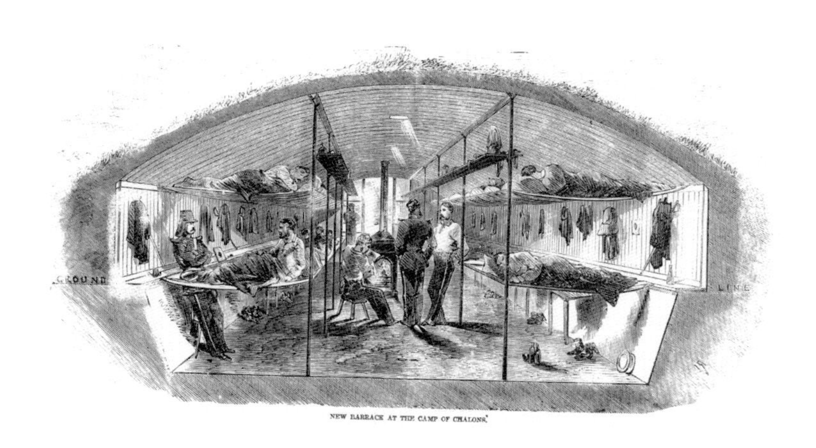

under £76,500, making him the third largest contractor for buildings in military encampments related to the Crimean conflict. By any measure, this was a spectacular expansion in business for a firm that had only been in existence from the early 1850s.

At least two other corrugated iron buildings were erected at the Aldershot Camp: a small prison, and a hut which *The Times'* reporter thought was 'probably intended for the General' (4 April 1855).

Tahurangi House, Mount Egmont (Taranaki), North Island, New Zealand, (pictured *c.*1930) is the sole surviving armoured 'iron house' from the Land Wars of the 1850s. Today the structure is used as overnight accommodation for climbers.

The multiplicity of uses suggests a degree of open-mindedness towards the material that was not replicated within broader society, and particularly the Anglican Church at the time (see chapter 3).

Another entrepreneur hoping to sell his wares to the military was Joseph Francis, an American engineer who had been experimenting with corrugated iron hulls for at least a decade. In 1856 Colonel Joseph Portlock, Inspector of Studies at the Royal Military Academy, Woolwich, urged the British government to consider the use of Francis' invention: '... however habitually cautious [the government] might be in admitting great military changes... it is hoped that it would follow the example of the United States and of Napoleon III by adopting in the army and navy the boats of Mr. Francis.' Six years later the boats were in use at the Woolwich Arsenal. '[They work] exceedingly well... the only drawback [is that] the pull is heavier than that of the ordinary wooden boat.' Francis appears to have found more success with his 'corrugated iron army wagons', which were designed for 'the transport of men and stores... where bridges may not be available'. A number were in use in naval dockyards during the 1860s.

During the middle of the nineteenth century corrugated iron was used in several colonial campaigns. A group of nine armoured 'iron houses' was built in 1856 for the protection of British troops during the Taranaki Land Wars in New Zealand. The long, barrel-vaulted structures were prefabricated in Melbourne, Australia – to a design devised by Captain Clarke of the Royal Engineers – and shipped over the Tasman Sea. In 1874, at the end of the conflict, they were used as temporary housing for immigrants until 1891, when eight of them were demolished. The ninth – now known as Tahurangi House – was sold and dragged by sled 1,000 feet up the north flank of Mount Egmont, where it is still used as overnight accommodation for climbers.

Māoris also recognized the merits of corrugated iron during the Land Wars. Indigenous New Zealanders are known to have used the material in the construction of 'pahs' (fortified encampments). For instance, in 1869, Tito Kowaru's defence of Nukumaru held up British forces for some weeks; it might have been much less without corrugated iron. When the invaders did finally break through, once the pah had been abandoned, they found, '... large underground chambers... [connected] by corrugated iron covered ways of great strength, the iron being plundered from the abandoned Waitotara homesteads'.

The Boer War

Isolated examples aside, corrugated iron was largely overlooked by the British military throughout the second half of the nineteenth century. Even during the Second Anglo-Boer War (1899–1902) the British sick and wounded were treated in wind-blown tents without drainage, no doubt contributing to the heavy losses.

But although it was not adapted for hutting, corrugated iron did have a role to play in the conflict. The barracks of the British military personnel running the concentration camp at Uitenhage, near Port Elizabeth were made of corrugated iron, as were Boer camps for British prisoners of war. It was also fundamental to a design innovation that hastened the end of the conflict.

A blockhouse is a small, isolated guard post. During the Boer War, the first were built in March 1900 to protect the British army's supply route from Cape Town to Bloemfontein. They were two-storey stone-built structures, linked by barbed-wire entanglements that cost at least £800 each and took three months to complete. Later in 1900, when the Boer guerrillas returned to the veldt, the challenge of protecting territorial gains and limiting the movement of the enemy became more complicated. The blockhouses and barbed wire had proved effective, but a quicker, cheaper version was needed.

Lord Kitchener, the new Commander-in-Chief, appointed Major S. R. Rice of the Royal Engineers to take up the challenge. His design for a prefabricated blockhouse was based on a corrugated iron prototype trialled to good effect in the Transvaal. It was a circular structure with a double skin of corrugated iron; the cavity was filled with rubble to make it bullet proof. It had a diameter of thirteen feet and a clear height of six feet under a gabled roof. When produced in large quantities, each blockhouse cost only £16. They could be erected by a team of six in six hours. The first line of Major Rice's blockhouses was built in January 1901 from Kaapmuiden to Komati Port. By the end of the war, seventeen months later, over 8,000 had been built, an average of forty per week. They formed the noose that strangled the Boers' resistance.

The First World War

Following the Boer War, the British search for efficient hutting moved up a gear. The limitations of tents were now widely recognized, as were the advances made in the construction of temporary barracks by other nations. But there was still no clear

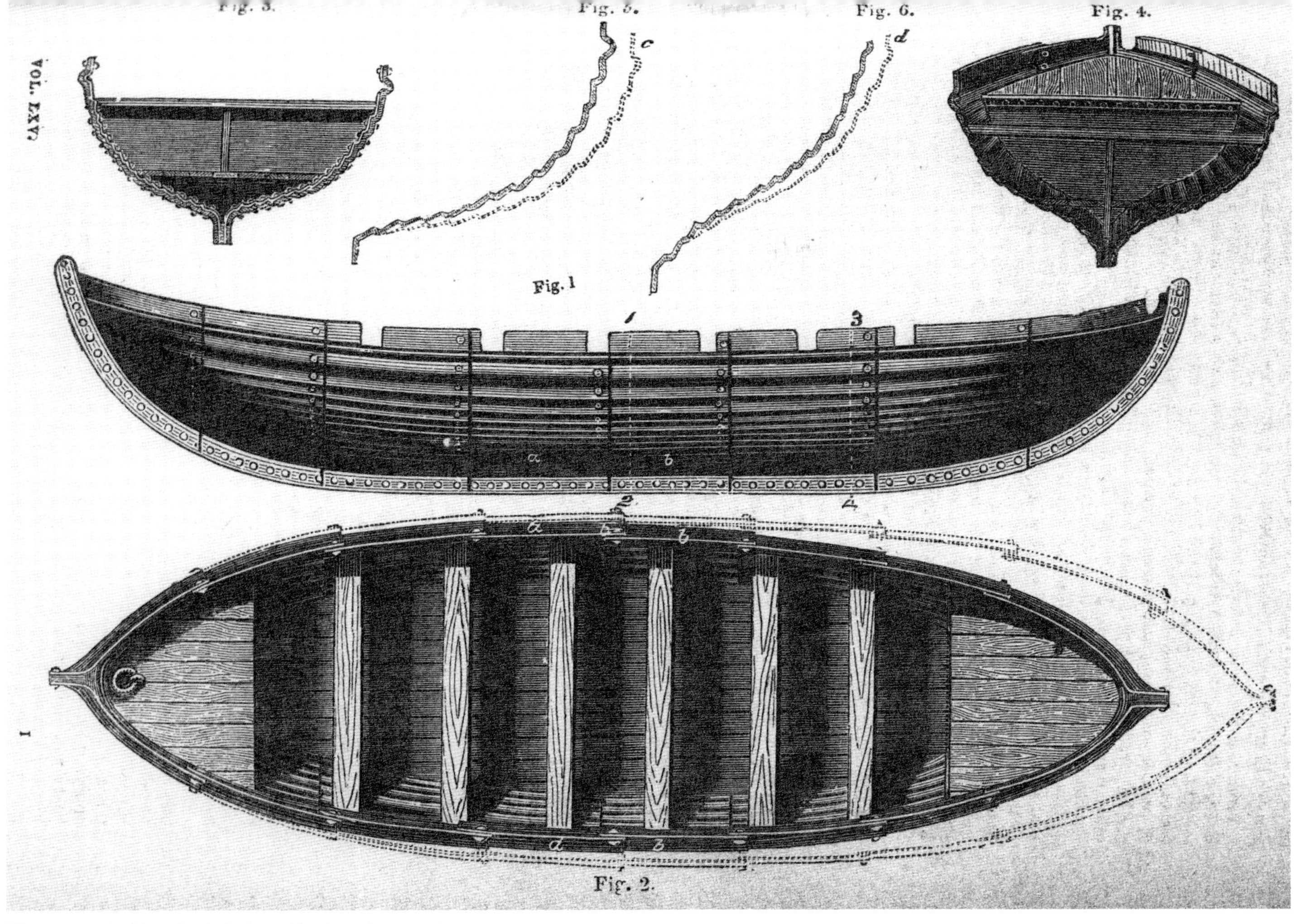

The corrugated iron boat designed by American engineer Joseph Francis proved popular with the French and American armies, less so with the British.

One of 8,000 corrugated iron blockhouses built in southern Africa between January 1901 and May 1902. Connected by barbed wire, the fortified guard posts enclosed a vast area of the veldt, and played a significant role in strangling the Boers' resistance.

BELOW Peter Nissen's design capitalized on the inherent qualities of corrugated iron to create a simple to transport, easy to build landmark of mass-produced portable design. (IWM 40791)

RIGHT The 'Military or Navvy's Hut' designed by Liverpool manufacturer Isaac Dixon in 1885 bears a striking resemblance to Peter Nissen's hut designed 30 years later. It is not known whether it was ever built.

OPPOSITE During the First World War corrugated iron was vital to Captain A. R. Turner, Chief Engineer of the British East African Pioneers. It gave him shelter, allowed him to store water, protect roads and riverbanks from erosion and create reservoirs. (TOP, IWM HU64201 and BOTTOM, IWM HU64203)

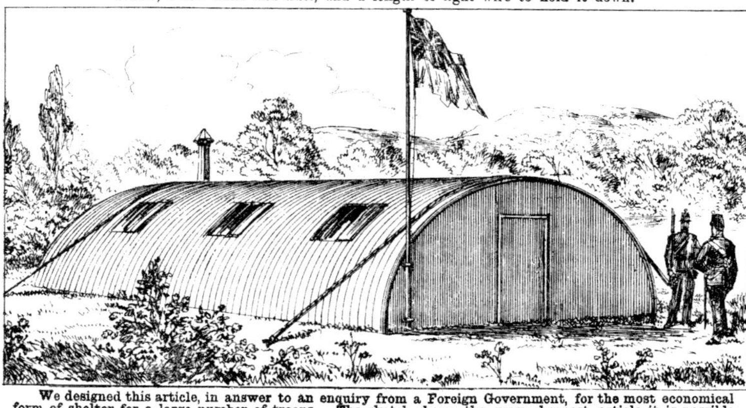

answer to the challenge of designing portable accommodation that could be built quickly by unskilled labour. During the early years of the twentieth century a number of ideas were trialled.

One was Armstrong Sectional Hutting, a system developed in 1912 for the construction of the Royal Flying Corps School at Upavon on Salisbury Plain. The huts – designed by Major B. H. Armstrong of the War Office – allowed the team of builders to live on site, which made the construction of the school possible in only six months. Major Armstrong noted that this was an improvement on the 'leisurely trundle of building operations normal in this country', but acknowledged that the huts were too bulky to be portable.

War with Germany was declared on 4 August 1914. Eight days later Major Armstrong ordered the deployment of Aylwin huts to France, but their lightweight frames with a canvas covering were poorly suited to the north European climate. Over the next two years a succession of huts were tried, each named after its inventor – Laing, Tarrant, Liddell, Weblee. To varying degrees, each was characterized by an adherence to conventional building forms, with walls, eaves and even foundations. They did not lend themselves to mass production or rapid erection in the field by unskilled labour.

The solution, a radically different type of structure, was found at the midway point of the First World War by Peter Norman Nissen, a Canadian mining engineer (see pages 118–23). Effectively half a cylinder of corrugated iron, the Nissen hut bore no resemblance to any hut that had previously been used in the field of action, although 60 years earlier a semicircular corrugated iron shelter had been designed for export to the Cape of Good Hope, and in 1885 a 'Military or Navvy's Hut' of strikingly similar form was included in a catalogue produced by Isaac Dixon & Co. of Liverpool.

The Nissen hut may have a claim to be the most replicated building ever designed. What is certain is that for over sixty years it played a major role in improving the comfort of military camps, and it remains an icon of portable architecture.

It was not just on the French battlefields that corrugated iron made its mark during the First World War. The material was also used in East Africa, the 'forgotten front', where British and German troops fought over colonial possessions, predominantly in German East Africa. Captain A. R. Turner, a surveyor and engineer, had volunteered in 1914. He served first with the Royal Engineers, and then as the Chief Engineer of the East African Pioneers. While in Africa his responsibilities included keeping road and railway links open and protecting water supplies. Corrugated iron was fundamental to his ability to do his job. For several months in the first half of 1916 Captain Turner was stationed at Longido camp, fifty miles north-west of Mount Kilimanjaro, close to the border with Kenya. It was a British stronghold in the campaign to oust German forces from the Kilimanjaro district, the most fertile corner of the German territory.

As well as living in a corrugated cottage, raised off the ground on timber posts with a broad verandah, he oversaw the construction and maintenance of countless iron water tanks and used the material to line a reservoir in a high rocky outcrop. Later in the campaign he used sheets of corrugated iron to shore up

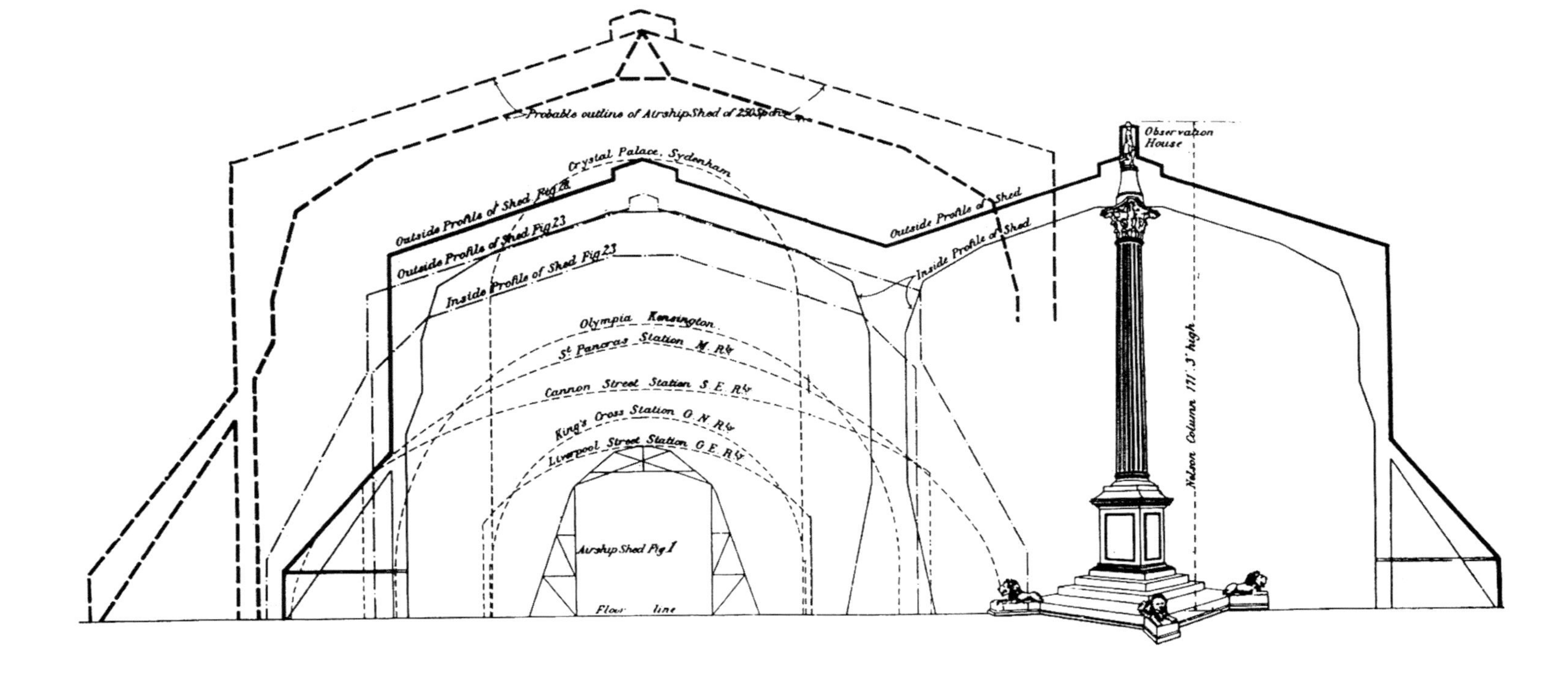

the banks of the Mgeta River, to try and save a suspension bridge after heavy rains. A resourceful and resilient man, Captain Turner was mentioned in dispatches in November 1916 and awarded the Military Cross in February 1919.

ABOVE An illustration *c.*1921 emphasizes the epic scale of airship hangars, structural behemoths that would not have been possible without the lightweight coverage provided by corrugated iron. The dotted line represents the anticipated scale of the next generation of airship hangar; the unbroken line represents the outline of the largest built shed.

OPPOSITE Bickendorf Aerodrome, Cologne, Germany, *c.*1919. The J1, an early incarnation of the famous Junker aircraft, had wings of corrugated steel. Later versions were made of duralumin, a lighter and stronger metal alloy.

The age of aviation

Almost as soon as air balloons, airships and airplanes started to play a role in warfare, corrugated sheet steel was harnessed to construct, house, launch and land these pioneers of aeronautical combat. The design outcomes included large-span industrial sheds, colossal airship hangars, even airplanes.

One of the world's oldest buildings continuously associated with aviation, and possibly the first to be associated with powered flight, is the Q3 building at the Royal Aircraft Establishment in Farnborough, England. It was originally built at Aldershot army base (*c.*1893) as the headquarters of the Royal Engineers Balloon School and Factory. It was moved in about 1906 because Farnborough was recommended as a better place from which to launch airships. In principle and form, Q3 set the tone for the steel-framed sheds sheathed in corrugated iron that defined the early years of large-span structures built for the nascent aviation industry.

Another surviving aircraft hangar at Farnborough is building R51 (*c.*1911), which was intended to be portable. The metal frame, which originally had a curved roof to accommodate airships, was covered in canvas. When war broke out R51 was reassigned as the Royal Aircraft Factory's foundry. Corrugated iron sheeting replaced the original canvas. The building still stands.

Also at the former Farnborough air base is the G29 building, also known as the 'Black Sheds', a hangar built in 1912 for the newly formed Royal Flying Corps, the precursor to the Royal Air Force – whose headquarters at Upavon on Salisbury Plain was built in the same year (see page 112). The shed became the pattern used throughout the Great War. It is a very rare survivor.

Another chapter in the early history of corrugated iron and military aviation was set on England's south coast. The former flying boat hangar at Calshot in Hampshire (*c.*1914) was built to accommodate the Sopwith Bat Boat. The 800-square-yard (720-square-metre) timber-framed shed, rectangular in plan, is clad entirely in corrugated metal. The 'Canoe' or 'Sopwith' hangar is just one of an outstanding group at Calshot. The largest, known as the Main Hangar, was built in 1918. It also accommodated flying boats. Its steel frame is supported by a lattice of steel girders and is today clad in corrugated asbestos, although corrugated metal would almost certainly have been the original covering material.

Airship hangars are, by some distance, the most remarkable buildings associated with the early years of flight. During the first thirty years of the twentieth century, German, British, French, Italian, Russian and American engineers were involved in the design of some of the largest structures ever built; one study of airship sheds describes them as, 'the most ambitious structures to be built since the Gothic cathedrals'. These lightweight, steel-framed hangars were absolutely colossal. Two surviving sheds at Cardington in England (see pages 130–31) could have accommodated *RMS Titanic*, save for forty feet of her bows; each of the shed's 470-tonne doors takes fifteen minutes to open and close; four tonnes of paint is required to paint each of the sheds; Nelson's Column could stand inside them. The statistics are endless.

Given their scale, it is remarkable that the engineers behind these structures actually tried to make them manoeuvrable. This was desirable because airships – long, cigar-shaped vessels – were difficult to handle, especially in high winds. Creating

hangars whose longitudinal axis was parallel with the direction of the wind would make it easier to house them safely. German engineers were at the cutting edge of these innovations. Ideas included designing hangars on floating pontoons and turntables on dry land. Both tasks presented an enormous challenge, and would have been almost inconceivable without corrugated sheet metal – any other material would have been too heavy.

In 1909 British engineers – who were well behind their German counterparts, the leaders in the field – attempted to design an airship shelter with 'temporarily removable roofs... to facilitate ingress or egress in a vertical direction... the roof covering is of a flexible corrugated metal capable of being let down behind the side walls, which are also of corrugated metal supported by braced pillars'.

Sadly, the era of airships was short-lived. They were an expensive failure, whose decline – precipitated by the Hindenburg crash in 1937 – coincided with the rise of the airplane. The hangars that survive, from Recife, Brazil to Bedfordshire, England stand testament to the first attempt to create a worldwide air service.

Another notable episode in the history of corrugated sheet metal and military aviation were Junker airplanes, developed from 1914 by German engineer Hugo Junker. During the First World War the 'J1' flew with wings of corrugated steel. One of Junker's later products, the J3, was the world's first airplane with an all-metal fuselage; it was sheathed almost entirely in corrugated duralumin, a lightweight metal alloy. The material was lighter than steel and stronger than canvas. Throughout the inter-war period, corrugated metal sheeting became the defining feature of the Junker's brand.

Perhaps the most famous of all the Junker aircraft was the Junker Trimotor (JU-52), nicknamed *Tante Ju* (Aunt Junker). It was a commercial aircraft sold in at least twenty countries during the inter-war years, from Ecuador to Estonia. During the Second World War the Luftwaffe used it for the transport of troops and equipment. The enduring appeal of the duralumin Junker is evident in the sale of watches with a miniature corrugated dial face.

Second World War

After their success in the First World War, the 1939–1945 conflict gave the Nissen hut a new lease of life. The original huts were built with timber ends, but a shortage of wood necessitated changes. Many of the huts erected during the Second World War had concrete floors and brick ends, which cut short any lingering

Alan and Doris Suter clamber into an Anderson shelter in their garden on Edgeworth Road, Eltham, south-east London. (IWM D778)

hopes that the huts could be portable – few Nissen huts were ever dismantled and rebuilt for military purpose.

As the conflict progressed and iron fell into short supply, asbestos, concrete and plasterboard were tried as alterative roofing materials. All were flawed. The asbestos and concrete failed to stand up to rough handling and the plasterboard leaked.

There were also experiments with new spatial variations. The Romney Hut, introduced at the end of 1941, was essentially an enlarged and improved version of the original Nissen Bow Hut. The semicircular shape was retained, but with a stronger frame: two inch-thick tubular steel ribs replaced the T shaped originals and the wooden purlins were discarded in favour of one and a half inch angle iron. The Romney measured 96 feet by 35 feet (the Bow Hut was 27 by 16) and saw service in the North African desert campaigns.

The Second World War also saw the introduction of at least two new types of corrugated metal kit buildings, the Quonset hut, America's interpretation of the Nissen hut (see pages 124–29) and the Anderson shelter, the quick-fix solution to the Luftwaffe's anticipated air strikes on industrial Britain.

In November 1938 John Anderson, then Lord Privy Seal, was placed in charge of air raid precautions. He commissioned engineer William Paterson to develop a cheap shelter that could be erected in private gardens and accommodate a family of six. Their design was based around fourteen sheets of corrugated iron. The main body of the shelter was composed of six curved sheets bolted together at the top. Additional plates were installed at either end. The shelters were 6 feet high, 4 feet 6 inches wide, and 6 feet 6 inches long, and were buried 4 feet in the ground. Soil to a minimum depth of 15 inches was used to cover the roof. The shelters cost £7 each, but were free to anyone earning less than £5 a week. By the time war broke out in September 1939 there were two million in gardens around the country. The shelters proved relatively successful but were not particularly popular. They were damp, dark and prone to flooding.

In March 1941 Home Secretary Herbert Morrison introduced an alternative shelter. They were effectively cages made of very heavy steel that doubled as dining tables. The idea was that they would protect a family as their house collapsed around them, allowing them to crawl out unharmed once the danger had passed. They were warmer and drier than the Anderson shelters, if not necessarily safer – if a house was on fire, or the fallen rubble inconveniently placed, the occupants were trapped.

During the Second World War two basic approaches to the use of corrugated sheet metal matured. One was as part of a closed system of construction as typified by the Nissen hut, where curved sheets of a particular radius married up with other components to make up predictable and unvarying structures. The other was for more spontaneous design ideas, many of which were tried very successfully, harnessing corrugated iron's almost infinite adaptability.

For instance, in British cities barriers made of corrugated iron were erected around landmarks to protect them from bomb damage, and at the Battle of Keren in Ethiopia (March 1941), an Indian brigade used sheets of corrugated iron as shields against grenades thrown from above by Italian forces defending the high ground. It was an unconventional battle – the terrain was so hard that digging trenches was not possible. Steel shortages in the Pacific also forced a number of innovations in the design of large-span, timber-framed aircraft sheds at American 'Forward Bases' in Australia and later throughout the Pacific Rim. These large arched structures relied on daring engineering, short lengths of timber simply nailed together, unskilled labour

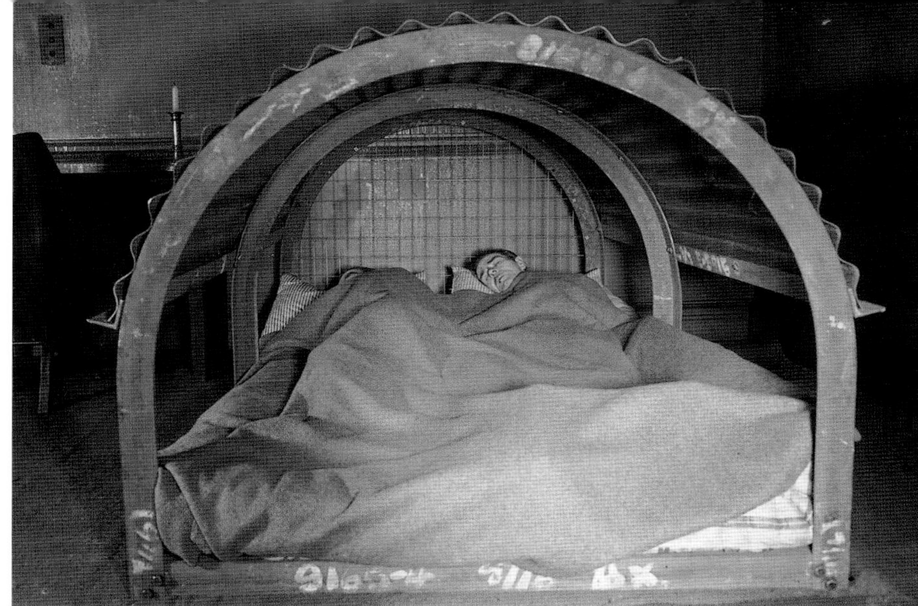

ABOVE LEFT Latham Street, Poplar, East London: A bomb landed a few yards away from the Anderson shelter. Three occupants survived the blast. (IWM D5949)

ABOVE RIGHT The double bed enclosed by corrugated iron was a short-lived innovation in protecting civilians from the impact of bombs. (IWM D2182)

LEFT Statue of Charles I, Trafalgar Square protected by a barrier of corrugated iron. (IWM D3606)

and the universal roofing material – corrugated iron. One was built with a spectacular span of 175 feet (55.2 metres) to house B-29 bomber aircraft in Darwin, but most were built to a 107 feet (32 metre) span. Some have survived.

In some respects, this more open approach to construction lent itself to the unpredictability of war, empowering people to do the best they could with the available supplies, the shifting priorities and the long lines of communication.

By the end of the war, corrugated iron was a well-established feature of military life around the world. It was used for shoring up trenches, providing shelter for troops and displaced civilians, and accommodating the construction and storage of aircraft among countless other planned and improvised applications. It has continued to be used in conflict zones ever since. The face of modern warfare would be quite different without it.

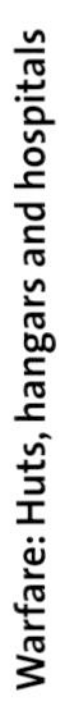

NISSEN HUTS

Her Majesty the Queen inspecting huts at Hesdin, Pas-de-Calais on 7 June 1917. Major Peter Nissen is on the queen's right. (IWM Q2529)

Peter Norman Nissen (1871–1930) was born in New York, brought up in Canada and spent much of his youth working with his father, a mining engineer. Like many young men during the late nineteenth century he was an enthusiastic gold prospector. He was also an inventor; during his relatively short life he registered 28 patents. The first, in July 1896, was for 'Pneumatic Boots and Shoes' (Canadian Patent No. 52953). The idea was to, 'increase the comfort in walking [by inserting] a rubber... cushion inflated with air... into the sole of a boot'. It did not go into production.

But 1896 was not an entirely fruitless year for Nissen. He also came across the building that would inspire the invention that made his name. While studying at Queen's University in Ontario he played ice hockey in the Drill Shed. He never forgot the large semicircular structure.

Nissen left university without a degree and dedicated the following decade to the pursuit of gold. He prospected in Utah, Arizona, Mexico and South Africa, with limited success. In 1912 he settled in Britain and two years later he joined the British army.

The story of how his experience as an engineer came to the fore may be apocryphal. Dismayed by his platoon's chaotic efforts to shift timber boarding during training, Nissen demonstrated a more efficient method that impressed an officer. 'You must be an engineer,' he said, to which Nissen replied, 'Of course I'm an engineer, but it doesn't seem to matter in this bloody war!' A few days later Nissen was transferred to the 103rd Field Company Royal Engineers and by the beginning of 1916 he was stationed at Ypres.

That year Verdun was besieged. A British and French offensive was planned for July to relieve German pressure. But because of intense shellfire over the previous two years, the nearby villages had been largely destroyed, so there were insufficient billets. The problem came to the attention of temporary lieutenant Nissen. Legend has it that the design came to him during a fitful night's sleep. He rose from his bed and began sketching a semicircular hut based on the old Drill Shed. The idea was put before the British engineer-in-chief. In March 1916 Nissen was transferred to General Headquarters where he began designing a prototype.

The fundamental features of the design were a semicircular form composed of curved corrugated iron sheets supported on a steel frame. Windows were installed in the timber walls at both ends. The structure sat on a timber base frame; the floor and internal lining were also of timber.

FAR LEFT AND LEFT To ensure that the huts were packed and transported efficiently, Peter Nissen prepared instructions describing the order in which the components should be put into lorries, leaving room for three men. (IWM H40790 and IWM H40793)

BELOW Some of the first Nissen huts ever built under construction at Fricourt, September 1916. In typical circumstances, a team of six men could erect a hut in four hours. (IWM Q1402)

ABOVE A road lined with Nissen huts used as billets near Mametz Wood, France. The photograph was taken on 2 September 1917. (IWM Q2783)

BELOW Huts at Fricourt in December 1916. (IWM Q1711)

The Nissen-Petren houses are a legacy of Nissen's attempt to adapt the hut to mass housing. It failed; only 24 were ever built, all of them in Somerset, southwest England.

The first hut was inspected in May 1916. After a number of revisions and some fine-tuning two standard models were agreed: the Nissen Bow Hut, which was 27 feet long, 16 feet wide and 8 feet high, and the Nissen Hospital Hut, which measured 60 feet by 20 feet wide and about 10 feet high. The first order was placed in August, by which time Nissen had been promoted to Major.

As well as being light and easy to transport, it was essential that the huts were simple to erect. Corporal Robert Donger, a mechanical draughtsman who in May 1916 became Nissen's lifelong right-hand man recalled: '[The Major] had in mind erection instructions which would be foolproof.' Annotated drawings showing each stage of construction in sequence were issued with every hut. To ensure that the huts were packed and transported efficiently, Nissen and Donger also prepared instructions describing the order in which the components should be put into the lorry, leaving room for three men. There was a diagram of every component, each of which had an index letter.

Within weeks firms across Britain were manufacturing components. John Summers & Co. and Braby & Co. produced corrugated sheeting; the Thames Joinery Company and Boulton & Paul of Norwich were employed to prepare timber panels; and William Baird & Co. of Coatbridge made steel ribs. Six months after the first order was received, over 20,000 huts had been put up on the French battlefields. That figure was 100,000 (accommodation for 2,400,000 men) by the end of the war. The final figures for hospital huts were 10,000 huts, providing space for 240,000 beds.

The speed and efficiency was impressive (Henry Ford had adopted systems of mass production only six years earlier). And the pace was matched on the ground. It typically took six men about four hours to erect a Nissen hut; the record was 1 hour 27 minutes.

Nissen's huts – popularly known as ''arf cartwheels' – were a great success. Well-organized hospitals and heated huts with cookers, drying rooms and baths were islands of civilization after the horrors of the battlefield. A journalist described their appearance to contemporary eyes:

> At about the same time as the tanks made their debut... another creature, almost equally primeval of aspect, began to appear in conquered areas. No one ever saw it on the move, or met it on the roads. It just appeared. Overnight you would see a blank space of ground. In the morning it would be occupied by an immense creature of the tortoise species... and when such a pioneer found that the situation was good and the land habitable, it would pass the word [and]... its fellow monsters would appear so that in a week... you would find a valley covered with them... The name of this creature is the Nissen hut. It is the solution to one of the many problems that every war presents. The problem here was to devise a cheap, portable dwelling, wherein men can live, warm and dry... The hut has no walls. It consists of a roof, ends and a floor. The roof is simply an arch of corrugated iron so that there are no eaves or gables to fit. Anyone can put it up, but four men can do it easily in four hours. The only tool required – a spanner – is supplied with it. The whole can be packed on an army wagon and its weight is two tons, but no single part or package is heavier than that which can be uploaded by two men.
> All parts are interchangeable – the roof is in 28 pieces – all the same. You arrange them in three nine-foot sheets with one corrugation overlap. You go on fitting them together, anyhow, in any order and when they are all used up you find the roof complete.
> The Nissen hut will not keep out shells, but its round back lends itself to the most artistic camouflage. It has already earned a high character as a useful, tractable, kindly domestic beast. (Filson Young, *Daily Mail*, 6 February 1917)

A Nissen hut converted into a stable near Kolkata, India. Countless huts still stand around the world, adapted for any number of uses.

Peter Nissen received the Distinguished Service Order and was mentioned in dispatches, but, as for so many returning soldiers, adapting to civilian life proved an uncomfortable process for Nissen and his hut. While he was serving in the British army he knew that he had no claim on government funds. But after the war he hoped for some financial recognition of his efforts. The War Office offered £500, but Nissen rejected it, regarding the figure as derisory. Soon afterwards he discovered that the British government had been selling Nissen huts in peacetime.

As an experienced inventor, Nissen understood the process of taking out patents to protect his work – with the permission of the British government he had patented the Bow Hut and Hospital Hut in June 1916 (No. 105468.16, for 'Improvements in and relating to Portable Buildings'). Permission was also granted to patent huts in ten other countries.

So after the war, when Nissen approached an American purchasing agent, he was surprised to find that the British government had already sold 1,800 of the huts left on the French battlefields to an American buyer, and had accepted an order for a further 1,000. Huts had also been sold to France and Belgium.

Negotiations between the British government and Nissen's legal representatives led to a payment of £10,000. For huts already sold to the United States, the secretary of state promised to 'use his best endeavours' to obtain £3 10s. for each bow hut, and £10 for each hospital hut. In 1920 a supplementary payment of £3,500 was made on condition that Nissen waive any future claim for royalties.

With the issue of payment resolved, Nissen turned his attention to adapting the Nissen hut to alternative uses. The first contract won by Nissen Ltd, a contracting building company, was for the construction of the National Wool Sheds near Hull. The structure, which covered almost ten acres, was composed of 18 sheds each 552 feet long and 40 feet wide with a total storage capacity of 234,000 bales of wool.

This contract was a valuable source of revenue to Nissen. In 1922 he was able to mortgage several adjoining fields at Rye House, Hoddesdon in Hertfordshire to build a manufacturing depot. His greatest hope was that the Nissen hut system could be adapted to the mass production of housing. A contract with Yeovil Borough Council in Somerset provided the opportunity to make the necessary modifications. Working with Messrs Petter & Warren, architects to the borough council, Nissen designed a semi-detached house with roofs of corrugated asbestos-steel, supported by a framework of semicircular steel ribs. The first two Nissen-Petren houses were built in Yeovil in 1925. A further 22 were built in three villages near Yeovil – South Petherton, Barwick and West Camel.

The Nissen-Petren houses were Yeovil Borough Council's admirably ambitious bid to tackle a serious housing shortage – in the mid-1920s it had a housing list of around 600. Sadly, the sums did not add up. The Nissen-Petren houses cost £350 each, about £100 more than traditional houses. The plan to build more was abandoned. They remain an anomaly in the Somerset landscape.

Nissen's next contract was to supply huts to the Spanish army in Morocco, which between 1921 and 1926 waged a campaign against a separatist army led by Abdelkrim Mohamed al-Khatabi in the Rif Mountains. But in truth work was difficult to come by. Nissen Buildings Ltd survived its founder's death in 1930 and the company was given a boost by the Second World War, but in the early 1960s the Hoddesdon factory was sold to a furniture manufacturer.

Nissen huts were used in military campaigns for over 66 years, from the Somme to the Falklands. Their post-war and civilian adaptations mean that they live on today, all over the world.

QUONSET HUTS

BELOW Abandoned Quonset huts at Wendover Army Air Base, Wendover, Utah. Some were buried for added insulation and to disguise them from the air; others were left to the elements.

OPPOSITE TOP Two 40ft x 200ft Quonset huts at Barbers Point Naval Air Station, Honolulu. From 1942, corrugated steel sheets were attached horizontally, recalling Isambard Kingdom Brunel's use of corrugated iron at Paddington Station (see pages 24-27).

OPPOSITE BOTTOM Interior of a Quonset hut at Barbers Point. Like Nissen huts, the Quonsets were supported on a semi-circular framework of T-section ribs sheathed in curved and corrugated steel sheets.

The Quonset hut, the American version of the Nissen, was designed in a matter of weeks during April and May 1941. It was developed for the same reason as its British cousin: there was an urgent need for easy-to-erect, portable accommodation for military personnel and equipment that could be quickly mass produced. It also took the same semicircular form, and was sheathed in the same material: curved and corrugated steel sheets. During the early stages of development, it was even referred to as the Nissen hut. But as time passed, and the design was refined, notable differences emerged.

The origins of the Quonset hut can be traced to the Lend Lease Act of 1941. The Act empowered President Roosevelt to, 'sell, transfer title to, lend, lease or otherwise dispose of [articles of defence to] the government of any country whose defence the President deems vital to the defence of the United States'. Its effects were twofold: it cleared the way for America's allies to obtain much-needed supplies and it enabled the United States to acquire bases abroad to prepare for its involvement in the Second World War. Local labour and material resources were limited at these new Forward Bases. So in March 1941 work began on a prefabricated hut system.

The commission was given to George A. Fuller and Co., a major contractor already working on the construction of a number of American military bases, including a new naval facility at Quonset Point, Rhode Island. At the time there were no adequate facilities available for the production of the huts at Quonset Point so, together with the navy, Fuller selected a nearby site along the railway lines at a place called West Davisville. Materials could arrive by rail and finished huts could be shipped from the new pier at Quonset Point – huts for sites in the United States were dispatched by rail.

While the factory was under construction, Fullers appointed an in-house team to work on the design of the hut. Architect Otto Brandenberger was the team leader, with support from Robert McDonnell, Tomasino Secondino and Dominic Urgo. Their brief included two key provisos: the huts were to be arched, for strength and to deflect shrapnel, and quick to erect and demount.

The team worked fast. The first known drawings were issued on 4th April. On 10 April the Bureau of Yards and Docks authorized an order of 2,488 American 'Nissen' huts for the bases in Scotland and Northern Ireland. On 11 June, when the *Empire Gull* embarked from Quonset Point, its cargo included the first run of huts.

The first hut, which came to be known as the T-Rib Quonset, was produced with two footprints – 16 feet by 20 feet, and 16 feet by 36 feet. It was similar to the Nissen in most key respects, although there was one notable area of improvement: insulation. The American huts were lined with lightweight compressed timber and a layer of wadded paper insulation.

128

PREVIOUS PAGE Three Quonset huts corroding in Unalaska, the Aleutian Island *c.*1985, a legacy of the Second World War American Naval Operating Base.

OPPOSITE TOP AND BOTTOM Thousands of Quonset huts have found civilian uses, including the Rio Theater, Monte Rio, California, and the Cinadome Theater, in Hawthorne, Nevada.

LEFT A Quonset hut in Nevada reborn as a welding workshop.

It was not until July 1941 that the huts were given their own name. Up to that point they had been referred to as 'Standard' or 'Nissen' huts. That changed when Lieutenant Commander E. Huntington, chief of the Bureau of Yards and Docks insisted that the term 'Quonset Hut' be used, 'in view of existing patents'.

Quonset huts were an immediate popular success. They were certainly a vast improvement on tents, the most common alternative troop accommodation. But work remained to be done. Given the climatic variations of American sovereign territory, from the cold climate of Alaska to the desert conditions of its western seaboard, and the increasingly global scale of the Second World War, it quickly became clear a one-size-fits-all hut was never going to be possible.

Throughout 1941 a number of improvements were made to the original Quonset concept. For warmer climates huts were designed with improved ventilation systems; later manuals included the option of overhangs created by inset bulkheads. Quonsets were also raised off the ground on timber stilts to allow air to circulate and reduce the threat of termite damage. In cold climates a common field modification was the addition of separate enclosed entrances that stopped cold air from entering the hut itself.

Different functions also demanded different design solutions, foundations and internal arrangements. In total, over forty types of hut were designed, including a hospital hut, laundry, latrine, dentist, pharmacy, barber, tailor and a morgue.

During the first winter of deployment, the navy also discovered that when square-backed furniture was placed against the sides, a significant amount of space was lost to the curve of the hut. The solution, developed towards the end of 1941, was an arch supported on two four-foot vertical sides. Another improvement was a new structural system, designed by Stran-Steel, whose lightweight steel frames had been developed for use in residential applications. The ribs of the arch were created by tack-welding two 'C'-shaped channels back-to-back. The gap between the channels was serpentine in shape, which meant that nails driven through the outer layer of corrugated steel sheets into the cavity were held by friction. It also meant that once the frame was complete, only a hammer was required to erect the huts.

By the time of the Japanese attack on Pearl Harbor, on 7 December 1941, the new and improved Quonset was on the production line. Within weeks demand had tripled. The Rhode Island plant was quickly overwhelmed. From the middle of 1942 the navy contracted responsibility for production to Stran-Steel, whose system had become intrinsic to the success of the huts. As a result, Detroit became the centre of production.

One of Stran-Steel's innovations was a new method of cladding in which corrugated sheets were oriented horizontally along the sides, as opposed to running vertically over the arch. Factory-curved sheets, with corrugations running in the opposite direction, were only used at the apex of the curve. It was a pragmatic decision: the predominant horizontal sheets did not need to be bent and therefore took up less space in transit. The only drawback was that the horizontal corrugations were not so effective at draining rainwater and snow.

It is not known exactly how many Quonset huts were built during the Second World War – estimates start at around 150,000. The number sold to civilians after the war, to help address the chronic housing shortage, is even more difficult to estimate. Some huts were sold second hand typically for around US$1,000, and Stran-Steel continued producing new huts until the late-1950s.

Today thousands remain across the United States, converted for any number of uses, including cinemas, homes, stables, workshops and churches. In some areas Quonset huts have become much-loved features of the landscape. Once again, the parallels with the Nissen hut are unmistakable.

LEFT The R101, or 'socialist airship', at Cardington in 1929. Each of the shed's 470-tonne doors takes 15 minutes to open.

BELOW During the Second World War, Cardington became the Royal Air Force's principal barrage balloon operations training centre.

THE CARDINGTON HANGARS

Bedfordshire, England

German Zeppelins began their air strikes over Britain during September 1916. That autumn, intelligence revealed the imminent threat of new Super Zeppelins. The Cardington hangars, remarkable survivors from the age of airships, were fundamental to the British counter-offensive.

In late 1916 the Admiralty bought land at Cardington and commissioned Short Brothers to develop an airship to respond to the German threat. It was an ambitious undertaking. Despite having experimented with the idea since 1908, the British had still not succeeded in getting an airship into the air.

Claude Lipscomb led the design team. They worked from plans provided by a Swiss engineer, Herr Muller, who had worked for the German airship builder Schütte-Lanz Luftschiffbau. Together they planned a vessel 615 feet long by 65 1/2 feet in diameter and a hangar in which to build and house it. Building 80, as it was known, was completed the following year.

Wartime restrictions required an Act of Parliament to release the 4,000 tons of steel needed for the construction. The hangar was 700 feet long by 254 feet high and 145 feet wide. It included 25 bays of steel framing with side aisles for workshop annexes and a huge central nave. Six steel stairways led up to three elevated catwalks. At the west end two enormous electrically operated doors opened to reveal the entire height and width of the nave – the airship construction area. The whole was sheathed in galvanized corrugated steel sheeting 0.048 inches (1.22mm) thick. An area equivalent to 30 acres was required to clad what was then the largest hangar ever built.

Airship R31 was the first to be built there. Designed and built by Lipscomb's team, it was ready in just over two years. This was an impressive but poorly timed achievement – it was only five days before the Armistice of 11 November 1918. The R38 was subsequently built at Cardington, but the end of the war, and ensuing economic depression, saw military operations cease in 1921.

The 1924 announcement of the formation of the Imperial Airship Service saw a new era dawn at Cardington. The British government dreamed of an empire-wide travel service. Two new airships for state and commercial use were to be built at Cardington. The building of the R101 airship required Building 80 (also known as Cardington No. 1) to be extended. Four new bays were added in 1926 and 1927, extending the building by 112 feet. When this extension was completed another airship shed was relocated from Pulham in Norfolk to a site adjacent to No. 1 Cardington. No. 2 Shed, which was in broadly the same form as its neighbour, was to house the R100, an airship built at Howden in Yorkshire by the Airship Guarantee Company. It was delivered in December 1929.

An Act of Parliament was required to release the steel required to build the colossal hangars at Cardington, which today stand as rare survivors of the era of airships, a doomed attempt to create a worldwide air service.

The R100, built by a commercial company, was dubbed the 'capitalist' ship. The R101, the airship commissioned by Ramsay McDonald, Britain's first Labour prime minister, was named the 'socialist' ship. Neither airship lasted long. After the R101 disaster of October 1930, when the airship crashed in France on its maiden flight to India (its 48 dead included Sir Sefton Branker, the then Secretary of State for Air) the British government terminated its support for the airship programme. The R100, which had made a successful maiden return flight to Canada, was broken up and sold for scrap.

Cardington's fortunes revived after the formation of Balloon Command in November 1938, when it became the Royal Air Force's principal barrage balloon operations training centre, resurrecting the First World War national defence system of employing barrage balloons as a deterrent to German bombers. This was to be achieved through the development of thousands of kite balloons. It was a complex operation: every balloon had to carry two miles of steel cable, which required a trained crew who could monitor the balloon around the clock.

Since the end of the Second World War the enormous sheds have been put to good use by, among others, a manufacturer of smaller airships, researchers into fireproofing buildings, a fireworks store and the film industry.

5 INFORMAL COMMUNITIES AND DISASTER RELIEF

By the end of the Second World War corrugated sheet metal had been the material of preference in agriculture, industry and temporary buildings for over a century. But for the next three decades the material lost its position of preeminence in these and other markets. These were corrugated iron's wilderness years.

The problem was caused by the emergence of new products that performed similar tasks just as effectively, if not more so. Enormous post-war rebuilding programmes, particularly in Europe and Japan, and the demand for up to 16 million new homes in the United States, not only inspired industries in the construction sector to explore new materials, but also to look at new systems of production and marketing. In this environment architects talked of 'adapting wartime production techniques for civilian purposes', and manufacturers emphasized 'improved insulation properties', 'structural efficiencies' and 'corrosion resistance' when promoting their products. Corrugated steel seemed crude and old-fashioned by comparison.

But the material did not disappear. Instead its virtues of portability, endurance and versatility have seen it flourish in markets created by post-war shifts in the geo-political landscape. Since the end of the Second World War corrugated sheet metal has become one of the defining features of the informal communities, shanty towns and squatter camps of the developing world. It has also become a staple of disaster relief, providing shelter to millions displaced by natural or humanitarian catastrophes. It might even be argued that corrugated sheet metal protects more people from the elements than any other material in the world.

In the years following the Second World War corrugated iron had two main rivals in the market for low-cost roof and wall coverings. One was steel sheeting with angular profiles, rather than the familiar wave-like design. Each profile had different attributes. Some could span longer distances than their sinusoidal cousins; others could be used to create decking to which insulation and waterproofing membranes could be attached. But, crucially, all of them looked 'new', and in doing so automatically appealed to fresh markets.

The other rival was not metallic, it was an entirely different product, and for a brief period it all but eclipsed corrugated sheet steel in the markets that it had once dominated. Asbestos is a fibrous mineral, whose insulating and fire-resistant properties have been known for millennia. The ancient Greeks used it. They also recognized that it causes lung disease.

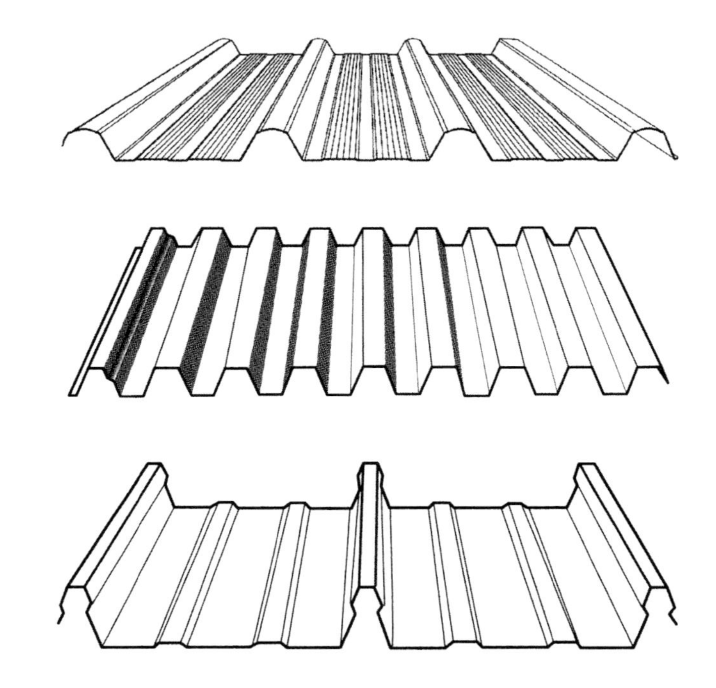

ABOVE Every angular profile has different attributes and a specific function.

OPPOSITE PAGE Miskito Nicaraguans being repatriated in 1988 after five years in Honduras refugee camps. UNHCR contributed sheets of corrugated metal to assist in the reconstruction of their homes in the village of Tasba Pain.

The modern revival of asbestos dates to the 1870s, but the greatest breakthrough occurred in 1900, when Austrian entrepreneur Ludwig Hatschek experimented with a combination of short asbestos fibres and cement to create a new type of building material. The sheets, branded 'Eternit', were 15 per cent asbestos and 85 per cent cement. By 1910 they were in production in at least ten countries. Corrugated sheets of the material were used in the First World War, and during the inter-war period they permeated all the markets that its metal equivalent had once dominated, their virtues of resistance to fire and corrosion seemingly outweighing their comparative weight and fragility.

The asbestos industry had a powerful marketing machine behind it: from the 1960s *AC* (Asbestos Cement), a Swiss periodical published quarterly with text in English, French and German, profiled technical advances and the latest products and accessories. Asbestos also received prominent public

OPPOSITE TOP Kibera, an informal community near Nairobi with a population of over 800,000, is a patchwork of recycled corrugated metal and other found materials.

OPPOSITE BOTTOM A school building in Kibera: public services and facilities are rare, a legacy of the community's status as an, 'administrative grey area'.

endorsements: the British government was one of several administrations that actively promoted the use of the material. From the 1950s until the 1980s a vast amount of asbestos was used in construction products and materials. Most buildings constructed or refurbished anywhere in the world during this period are likely to include some form of asbestos-containing material. The material seemed omnipotent, which made its breakneck fall from grace all the more dramatic.

Rumours of links between asbestos dust and lung disease – particularly mesothelioma, a malign tumour – had circulated for decades. A turning point came in 1969, when the first third party products liability suit claiming personal injury from asbestos was launched in the United States. The jury found in favour of the claimant. By 2000 the cost of asbestos litigation to American business was over US$50 billion.

Informal communities

In the decades following the Second World War asbestos cement may have usurped corrugated sheet metal in its established markets, but in the developing world it was a different matter. In informal communities on the peripheries of urban centres in Africa, Latin America and South Asia – not primary targets for the asbestos marketeers – corrugated metal's well-established virtues were still in great demand; its portability in comparison to heavy asbestos sheeting was particularly appealing. These settlements, whose explosive growth coincided with the glory years of the asbestos industry, were – and remain – intimately associated with corrugated metal.

The scale and speed of post-war urbanization in the developing world was extraordinary. Between 1960 and 1993, despite falling wages and rising unemployment, Third World urban growth rose by an average 3.8 per cent per annum. Today, there are an estimated 200,000 informal settlements on the planet, with populations ranging from a few hundred to more than a million. In some countries almost all urban dwellers live in marginal peri-urban communities.

The factors that contributed to the growth of informal settlements – typically, self-built housing developments in peripheral, often hazardous locations, in which low-income residents do not own title and rarely have access to services – were many, varied and complex. During the 1950s and 1960s notable catalysts behind large-scale rural-to-urban migrations in the developing world included the global agricultural depression, the fall of South American dictators and African decolonization, although local conditions varied enormously.

For instance, in colonial Kenya, the British administration denied the local population both the right to own urban land and the right of permanent residency in urban areas. In the mid-1950s, when these regulations were relaxed, the Mau Mau Uprising (1952–60) against British rule was causing unrest and large-scale displacement throughout the country. The combined effect of these and other events was the gravitation of large numbers of people to urban areas, and many of them congregated in communities like Kibera, a small informal settlement along the Kisumu rail line about six miles south-west of Nairobi.

In the mid-1950s Kibera was populated largely by demobilized Nubians and their extended families – a 1948 census revealed that it had a population of 3,085. The Sudanese had been loyal military servants of the British in East Africa since the late-nineteenth century. They had no legal claim to the land; their tenure was based on their status as former servants of the British Crown. But the British had made no long-term plans for Kibera. As a result it became an 'administrative grey area', which made it attractive to lower income families and people on the fringe of society. Within only a few years, Kibera's population had mushroomed to over 100,000, made up of people from all over Kenya. Today it has a population of around 800,000.

The phenomenon of unplanned urban growth was repeated with regional variations all over the world. After Algeria's seven-year colonial war with France ended in 1961, up to half of the rural population moved to urban areas. The population of Algiers tripled within two years. To many of the migrants, home was a self-built shack made of sheet metal in a *bidonville* on the urban fringe. (*Bidonville* is the Francophone name for a shanty town; it derives from *bidon*, which means can or drum.)

Disseminating corrugated sheet metal

By the 1960s corrugated steel had been a feature of most European colonies for over a century. The typical trajectory for a sheet was to be manufactured in the mother country, before being shipped to the colony as part of a temporary dwelling or for a specific function – perhaps religious, commercial or agricultural. Once they had served their original purpose, or been replaced, the sheets might begin an extended life in the local community,

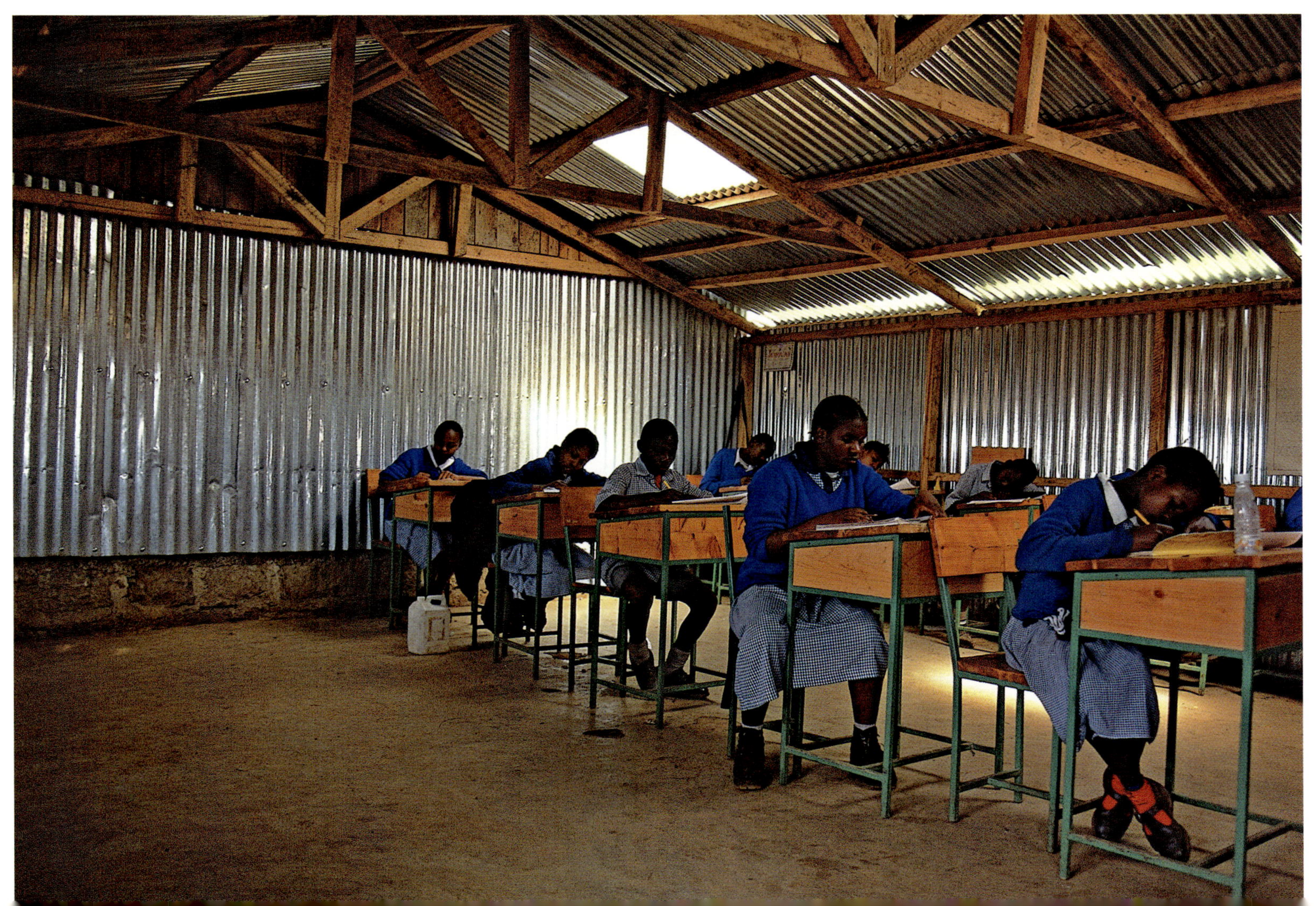

ABOVE The Red Location Museum and Cultural Precinct is located in the heart of Port Elizabeth's first settled black township.

OPPOSITE Twelve monolithic 'Memory Boxes' of rusting corrugated steel house exhibits offering different perspectives on the apartheid-era struggle for independence in South Africa.

where they would serve any number of uses, including the protection of crops, livestock and people.

By the end of the Second World War this process, allied to the increasingly widespread production of the material, meant that corrugated sheet metal had permeated most corners of the globe, even the small, isolated central African state of Rwanda.

In the mid-1990s, French author Jean Hatzfeld recorded that:

> Sheet metal arrived in Rwanda [with] the Belgians, after the First World War... it was intended as roofing for colonial buildings... With the passage of time and the emancipation of the people, [corrugated] sheet metal... spread into the towns, their outlying districts, and gradually the hills, to cover almost every dwelling, including the more modest, which have come to be known as terres-toles (sheet metal abodes).

In South Africa the process even inspired the name of a neighbourhood, as well as shaping its history and inspiring the form and identity of a bold new museum. The story dates back to the Second Anglo-Boer War (1899–1902), during which the staff barracks at the British concentration camp at Uitenhage were made of corrugated iron. After the conflict the barracks were relocated to nearby Port Elizabeth, where they became home to British soldiers. When the soldiers moved out, local black families moved in. As time passed, the rusting corrugated iron sheets began to stain the ground. The area became known as the 'Red Location'. It was Port Elizabeth's first settled black township.

During the apartheid era (1948–94), Red Location became a site of intense resistance; the suspended floors of the barracks were one of many places used for hiding people and weapons. Partly in recognition of the township's experience during the armed struggle, Red Location has been named a 'National Site of Struggle'. Since 2005 it has been home to the Red Location Museum and Cultural Precinct, designed by architects Jo Noero and Heinrich Wolff. The building draws on the area's history and physical form to challenge conventional ideas about museums, notably in the representation of history as a single narrative. At the Red Location Museum the past is represented as a set of disconnected memories, an idea given physical form in 'Memory Boxes', twelve large unmarked monoliths made of sheets of rusting corrugated steel, which were inspired by the boxes used by black workers to store their most precious possessions while they were forced to work away from home. Each houses an exhibit offering a different perspective on struggle in South Africa.

In some parts of the developing world the availability of sheet metal has contributed to the erosion of traditional building practices. The numerous examples include the inhabitants of the Kuiseb and Ugab watercourses of Namibia, for whom corrugated sheet metal has become a very practical alternative to traditional bark-covered houses – the sheets are fixed to a timber frame.

A resident of Klong Toei in Bangkok, Thailand. Informal settlements are in a constant state of evolution, adapting requires great skill and ingenuity.

To some extent this process of erosion is inevitable. In large-scale informal communities vegetable materials can be hard to come by on a mass scale, so corrugated metal becomes a viable alternative. The sheets are also fire resistant – a significant asset in tightly packed urban areas – and much easier to maintain than traditional materials. Portability is another important consideration. Residents of flood-prone informal communities have been known to keep corrugated metal roofs in place with rocks rather than nails, making them easier to remove in the event of danger.

Building informal communities

While local laws and regional characteristics influence the shape, location and scale of informal communities, one feature common to all of them is corrugated sheet metal. In some marginal communities the sheets are their defining feature. Typically, they are used for walls, roofs and barriers, often in combination with traditional building materials from local sources. They are also used to make paths over unstable ground, and can be used to channel water away from the sides of buildings. They are flexible, cheap and easy to work with and they can be recycled for a broader range of uses once their life as a wall or roof is finished. They might become part of a kitchen hut, toilet or animal enclosure, or find a new use as a boat or coffin.

The quality and age of the sheets vary enormously as Hatzfeld explains:

> [In Rwanda] the sheets last varying lengths of time depending on whether they have been imported from Europe (the best), Uganda, Kenya, or manufactured locally in the Tolirwa factories of Kigando, near Kigali. These homemade sheets are the thinnest and flimsiest. They last around 15 years.

The quality of informal housing itself is also highly variable, albeit underpinned by great ingenuity and resourcefulness. Life in an informal settlement is typified by flux. Most people exist beyond the rule of law, which means that efforts to consolidate and improve homes are typically carried out against a backdrop of insecurity. In such conditions design is a necessarily pragmatic process.

However, the physical arrangement of informal communities may reveal a human instinct for urban planning. Harnessing the topography of sites for drainage and sanitation, and arranging paths and alleyways to capitalize on solar gain are just two of the self-taught skills that make life in unplanned communities possible. In 1956, following its tenth conference, the Congrès Internationaux d'Architecture Moderne (CIAM) – an association of modernist architects and planners – actually praised the spontaneous order of the *bidonvilles* around Algiers for the 'organic' relationship between buildings and the site, which it described as reminiscent of the casbah.

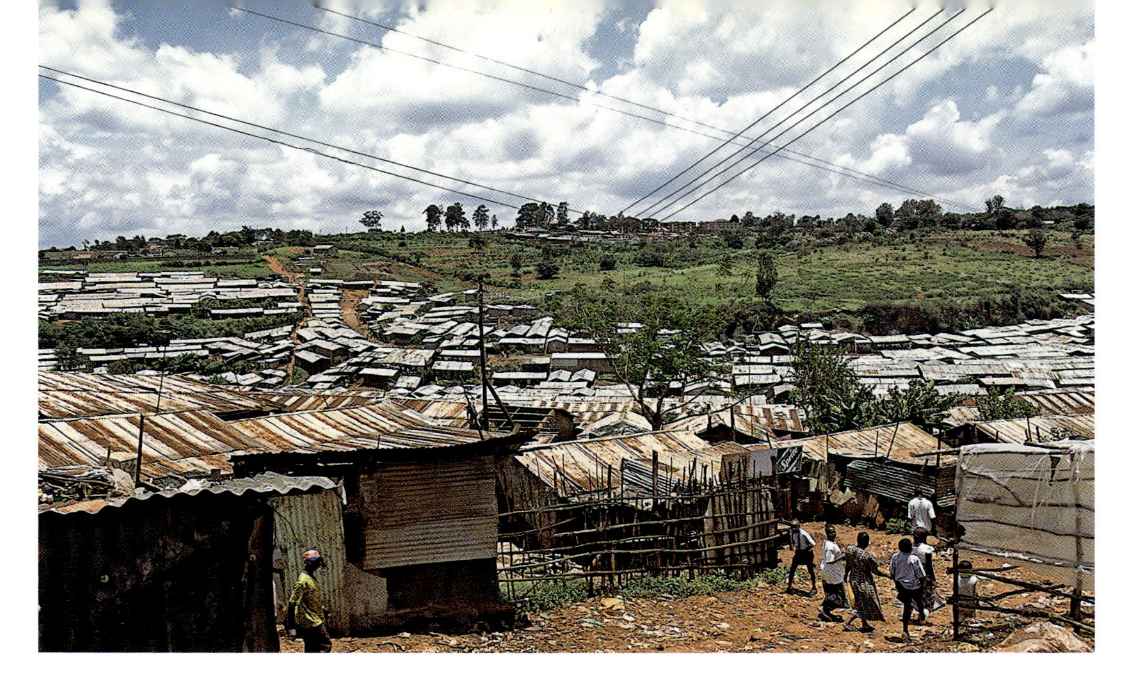

In Kibera, like all informal communities, recycling is a practical necessity. Nothing is wasted, including corrugated iron. Aside from the walls and roofs of dwellings, the material can be used for any number of purposes including fire drums, animal enclosures, even baths.

Sheets of corrugated metal work in tandem with mud and wicker to enclose self-built houses.

Without a communal civic commitment to sanitation and street cleaning life in informal communities would quickly become unbearable.

One of corrugated metal's greatest advantages over canvas or plastic sheeting is its capacity to carry heavy loads, vital in harsh winter conditions.

Corrugated currency

Of all the materials used to build an informal shelter, corrugated sheet metal is the only one that cannot be made by hand, which heightens its value considerably. In some Third World countries sheets of corrugated iron have literally become a currency. The standard price for a nanny goat in pre-genocide Rwanda was two sheets of corrugated sheet metal. A cow cost 20 sheets.

Prices rise and fall with the seasons; they are at their highest during the wet season, when the priority is to protect harvests from the elements. Civil war, social unrest and the potential for natural catastrophes are other factors that influence the price. If a family is forced to flee, their sheets of corrugated iron may be among their most valuable possessions. No matter where they are, with their rolled up sheets of corrugated metal on their shoulder, they can create shelter and have a trading asset.

In 1994 Hatzfeld watched thousands of Hutu refugees cross the border from Rwanda into Congo: 'Some carried a bundle or a child, while others lugged a chair, a basin or a sack of grain, and the strongest advanced bent double beneath... sheets of corrugated metal, which they exchanged for passage across the border, or a bag of grain.'

Many of the Hutus eventually found sanctuary in United Nations refugee camps in the Lake Kivu region of Congo. Goma was one of the largest; during the mid-1990s, it was home to a population of around 700,000 – of whom thousands died in a cholera outbreak. Many of the refugees lived in tents or under plastic sheeting, while others lived in temporary homes roofed with corrugated iron.

Disaster relief

In refugee camps, or any temporary settlement, corrugated sheet metal has a number of advantages over canvas and plastic sheeting, the other primary materials of humanitarian aid. Perhaps the most obvious asset is superior weather resistance. As well as being suited to temperate and hot climates – windows and extended eaves can help to moderate the temperature – metal sheeting can withstand monsoon rains and carry heavy snow loads. It is also possible to add layers of insulation; packed grass on the interior or rammed earth encasing the exterior walls are just two options.

Galvanized steel is also rot-proof, which means that large volumes can be stockpiled for long periods, and it does not deteriorate once in use. Since 2002 the Office of the United Nations High Commissioner for Refugees (UNHCR), the global

In Kashmir, following a major earthquake in October 2005, vast quantities of corrugated metal sheeting were rushed to isolated mountain passes before winter set in.

coordinating agency for disaster relief, has sought to address this problem by using a Lightweight Emergency Tent made of synthetic materials, but the problem of longevity remains: tents and plastic sheeting are only ever short-term solutions.

Of course there are situations where relief agencies want to encourage refugees to return home once conditions have improved. In such cases the message of semi-permanence projected by corrugated metal sheeting may not be appropriate. But in conditions where there is a high probability of the camp lasting for an extended period – perhaps due to an intractable conflict, or because a natural disaster has completely decimated the infrastructure of a region – it is a different matter. There are numerous examples of refugee camps morphing into permanent settlements, including Gaza, and Dadaab, a group of three camps in Kenya, which have been home to 125,000 Somalis displaced by the Somali civil war since the early 1990s.

In 'transitional' as opposed to 'temporary' communities, support agencies take a longer-term approach to the provision of shelter. Structures may be composed of corrugated roofing sheets, cement blocks and local timber, rather than plastic or canvas, and the occupants may take a hands-on role in the design and construction processes. This gives those displaced both a basic training, and some of the raw ingredients to build a permanent home in the future.

The principle of involving refugees in the design of shelters – also known as self-help building programmes – has been used since at least the early 1970s. Fred Cuny, an American disaster relief specialist, was among the first to use the built response to a disaster as the first step in the rehabilitation of the refugees' lives. In 1976, following an earthquake in Guatemala measuring 7.5 on the Richter scale, Cuny encouraged aid agencies to disperse corrugated steel sheets rather than tents. The thinking was threefold: first, the sheets could be used for the roofs of temporary shelters made of materials salvaged from damaged buildings; they could then be used in a more permanent dwelling; finally, once an alternative roof covering had been found, they could be traded for an alternative asset. For the same reasons, the principle of giving sheets of corrugated metal to repatriated refugees is now used widely by aid agencies.

The availability of materials also plays a part in determining the nature of temporary shelter following a disaster. During the last three months of 2005, following a major earthquake, vast quantities of corrugated sheet metal were taken into the isolated mountain passes of Pakistan-administered Kashmir. The reasons included: the material's load bearing qualities, vital to withstand

BELOW Bolivian schoolchildren learn about the role corrugated sheet metal can play in decontaminating water. Solar radiation is magnified by the reflection of the sun's rays off the galvanized metal.

RIGHT A contemporary rolling machine. Sheets of metal can be produced to almost any length and with a variety of profiles.

Expose the bottles to the sun morning to evening for at least 6 hours

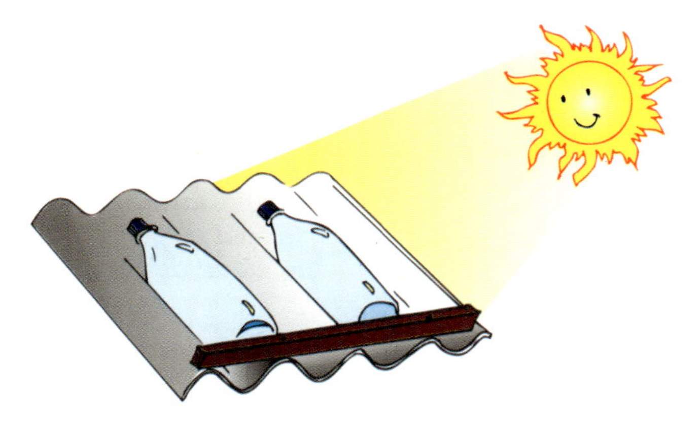

the mountain snows; the fact that metal-roofed shelters can accommodate whole families of different sizes; and its incombustibility, an important consideration given that aid agencies were also distributing kerosene oil stoves for warmth. Pakistan's status as one of the world's larger producers of corrugated sheet metal also played a fundamental role in its suitability for purpose. The vast majority of the sheets, and flat-headed nails required to attach them to the remains of buildings or improvised timber frames, were bought by overseas aid agencies from rolling mills in Pakistan. It was by far the quickest and cheapest way to get the raw construction materials into the mountains.

Disaster relief agencies, and organizations working to create self-reliant communities in a bid to prevent future disasters, form one of the world's largest markets for new sheets of corrugated metal. On a global scale only agriculture – arguably – is larger than the humanitarian aid market. And these agencies use the material in all manner of innovative ways. For instance, the Swiss Federal Institute for Aquatic Science and Technology, in promoting water decontamination through solar radiation, advises that plastic bottles filled with water should be left on corrugated iron roofs for at least six hours, until the pathogens have been destroyed. The corrugations support the bottles,

The quality and age of corrugated metal sheets used in informal settlements varies enormously. Some may have been imported, others rolled locally.

and the galvanized metal reflects the sunlight, so magnifying the solar radiation.

As well as confirming the continuing relevance of the material, the disaster relief 'industry' also reveals much about the nature of contemporary manufacturing practices. Perhaps above all, it reveals that production has moved east. Until the end of the nineteenth century Britain was the leading producer of corrugated iron and its successor sheet steel. During the mid-twentieth century American and Australian producers took over. Today China is by a distance the world leader in the production of rolls of sheet steel, and Turkey, Kenya and China are among the three most prolific consumers. In these and others countries rolls of un-corrugated galvanized steel are sold to small-scale industries with rolling machinery for sale to local markets.

Today, sheets of corrugated steel are a prominent feature of urban and rural landscapes throughout the developing world. A process that began with European colonization has been continued by the disaster relief agencies and the increasingly widespread production of the material. The ubiquity of the material is such that, even if production were to be terminated tomorrow, corrugated sheet metal would remain a feature of these landscapes for generations to come.

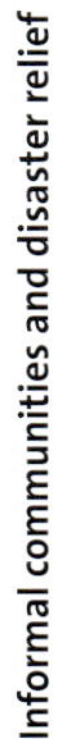

6 CORRUGATED IRON IN CONTEMPORARY ARCHITECTURE

By the 1950s corrugated iron had a rich and varied history. It had provided easy-to-build and affordable shelter from the tropics to the South Atlantic, and been used for applications as diverse as military hutting, lighthouses and missionary churches. It had also been embraced by leading engineers – including Isambard Kingdom Brunel and Gustave Eiffel – but architects had rarely paid it much attention, and it had never achieved widespread acceptance as a material of sophisticated urbane architecture. To many corrugated iron was still synonymous with poverty and impermanence; it was certainly not a material used by revered architects in the design of homes for millionaires. But from the 1950s that all began to change. For the first time since its novelty first began to wear off in the 1870s corrugated sheet metal could once again be described as 'modern', even 'fashionable'. In some parts of the world it still holds that status today.

The origins of corrugated iron's transformation into a component of fine architecture date to the late-1920s, when progressive architects and inventors began to consider ways of transferring manufacturing technologies from industry to architecture. Walter Gropius (1883–1969), a German architect and founder of the Bauhaus, was among the vanguard of innovators:

> Building, hitherto an essentially manual trade, is already in the course of a transformation into an organized industry. More and more work that used to be done on the scaffolding is now being carried out under factory conditions... and fabricated materials have been evolved which are superior to natural ones in accuracy and uniformity. (*The New Architecture and the Bauhaus*, Faber, 1935)

One outcome of Gropius' thinking was the Copper-Plate House, a five-bedroom home constructed entirely off-site.

Others working at the interface between industry and architecture included French architect and industrial designer Jean Prouvé (1901–84) – whose prodigious output included mass-produced aluminium shelters for export to tropical regions, and furniture of bent and compressed sheet metal – and the American engineer and polymath Buckminster Fuller (1895–1983).

Taking the premise that the home was an anomaly – a man-made product untouched by the Industrial Revolution – Fuller reconsidered every phase and component involved in its production. The result, first exhibited in 1929, was the Dymaxion House, a hexagonal building wrapped in aluminium sheeting – 'Dymaxion' was a composite of 'dynamic:maximum:tension'.

The essential structural difference between the Dymaxion House and conventional homes was the generation of strength through tension rather than compression. To achieve this, Fuller suspended the building from a central mast, which carried both its weight and replaceable units holding appliances and wiring. The building was decades ahead of its time: it could be lit with a single light bulb, thanks to carefully positioned mirrors; integrated wind turbines generated energy and rainwater was recycled.

However, the Dymaxion House did not go into production until 1941, and only then in a revised format. In collaboration with the Butler Manufacturing Company, a specialist in corrugating and pressed metal panel systems, Fuller developed the Dymaxion

OPPOSITE PAGE Julius Shulman's celebrated photograph of Case Study House no. 22 shows the gravity-defying structure sandwiched between a profiled steel roof and a sun lounger. A generation earlier these products would have sent out conflicting messages of destitution and indulgence, but by the 1960s perceptions of sheet metal were beginning to change.

BELOW In 1941 Buckminster Fuller collaborated with the Butler Manufacturing Company to produce a radically revised version of the Dymaxion House for mass housing. The few built examples were diverted to the war effort.

The roof of Albert Frey's second self-designed home, an elegant pavilion on an escarpment above Palm Springs, was made of corrugated aluminium, *c.*1964.

OPPOSITE Frey's first house in Palm Springs was sheathed in corrugated metal, applied vertically to the perimeter walls and horizontally on the wall planes extending into the landscape. The circular rooftop extension and swimming pool were added between 1948 and 1953.

Deployment Unit, a low-cost circular house in which curved and corrugated metal sheets functioned both as the structure and the building envelope. Arrangements were in place to produce 3,000 daily, but the United States government was unable to divert sufficient supplies of steel from the war effort to make it a reality. The American army used most of the few hundred circular Dymaxion houses that were built in overseas bases during the Second World War.

The generation of architects working in America during the 1940s and 1950s was among the first to accept corrugated sheet metal as an appropriate material for contemporary architecture. It was cheap, light and resonated with the enthusiasm for mass-produced machine-made components. On an aesthetic level, its shiny, streamlined appearance was suited to the times; it implied speed, flight and dynamism – a railway carriage patented by industrial designer Norman Bel Geddes was one of many products of the era finished in profiled metal.

From the 1960s young Australian and European architects also began to reassess corrugated sheet metal. In both regions, and in parts of South Africa and East Asia, it has been reinterpreted with great style and imagination. Architects have exploited its sculptural qualities (see Shuhei Endo pages 204–17), its potential to invoke industrial precision and lightness and its symbolic value in the development of regionally specific styles of architecture (see Glenn Murcutt, pages 180–91 and Lake|Flato Architects, page 192).

America – from the West Coast and beyond

In the American context corrugated iron has particularly strong associations with the Western states, especially California. During the gold rush of the mid-nineteenth century, the material was shipped there in substantial volumes (see Chapter 2); a century later, California became the source of new applications and attitudes towards the material. It remains a prominent feature of the landscape today. The mid-twentieth-century revival was inspired by a young generation of architects for whom California held numerous attractions, including its distance from an unsettled Europe and relative lack of entrenched urban traditions, especially when compared to the east coast. From the 1930s it also became home to new high-technology manufacturing industries, meaning that modern materials, and the skills to build with them, were readily accessible.

The small desert city of Palm Springs, at the base of the San Jacinto Mountains in the heat of the Colorado Desert, was a particular focus of architectural activity. From the late-1930s it

Case Study House no. 8 by Charles and Ray Eames included components from the Truscon Steel Company catalogue.

became known as the playground of Hollywood's rich and famous. It also became the home of 'desert modernism', a form of contemporary architecture developed in response to the harsh climate and arid landscape – sandstorms are common, and rain rare. One of the first architects to experiment with forms and materials appropriate to this climate was Albert Frey (1903–98), a Swiss émigré who had previously collaborated with Le Corbusier.

Among Frey's earliest buildings in Palm Springs was his own home, Frey House 1 (1941). The structure, rectangular in plan, was built around a series of horizontal wall planes that extend into the landscape, creating distinct outdoor 'rooms', an idea he continued to play with throughout his career. Another idea that became a feature of Frey's work was the use of machine-made construction materials, especially corrugated metal. He used the strong, resilient galvanized sheets to deflect the heat of the sun and power of the wind while simultaneously minimizing costs. Other examples of Frey's use of the material include a house for industrial designer Raymond Loewy in Palm Springs, the North Shore Yacht Club at the Salton Sea, and his second self-built home set on an escarpment above Palm Springs.

For much of his life Frey, like his fellow 'desert modernists' – including Richard Neutra, Marcel Breuer and John Lautner – was a fashionable, widely published and influential architect. There is little doubt that his endorsement of corrugated sheet metal played a small but significant part in the post-war popularization of the material.

Responses to the post-war housing crisis were also influential in changing attitudes to corrugated metal. For the first time machine-made metal components became widely acceptable for residential living, not just for industrial and agricultural applications. In the immediate post-war period the demand for housing was colossal. In the space of barely a few months, millions of military personnel returned to civilian life, and countless others went home as their wartime manufacturing jobs came to an end. A report ordered by President Roosevelt estimated that 3.5 million homes were needed in 1946 alone, but the building industry was ill-equipped to respond: it was still catching up with demand from the Great Depression and materials shortages caused by the war exacerbated the problem. Over a million families had been forced to 'double up' by the end of 1946; there were even reports of people visiting funeral parlours in a desperate bid to find recently vacated homes.

The extent of demand for housing had been anticipated before the end of the war. 'Designs for Postwar Living' a 1943 competition sponsored by California Arts and Architecture in collaboration with 22 material manufacturers, drew hundreds of entries. Finland-born architect Eero Saarinen and Oliver Lundquist, a naval veteran, designed the winning scheme, which drew on advances in the construction of metal buildings. The system was comprised of two standardized 'pre-assembled components' – one a living space, the other for all 'biological and mechanical' needs – which Saarinen and Lundquist hoped would be sufficiently flexible to address 80 per cent of America's post-war housing requirements (it was never built). Two years later, Saarinen used curved lightweight metal panels in the design of another mass-housing scheme (also unbuilt). This time the concept was based around a flexible aluminium roof that spanned between prefabricated metal living units.

Other leading figures in America's search for solutions to its housing shortage included husband and wife designers Charles and Ray Eames, and John Entenza, editor of *Art and Architecture*, a monthly magazine published in California. In 1944 the three of them, along with Buckminster Fuller and Swiss graphic designer Herbert Matter, collaborated on an issue of *Art and Architecture* dedicated to industrialization and mass production in the design of housing – it was published in July 1944. But it is for initiating the Case Study House program that Entenza and his magazine are best remembered.

Entenza launched the program in January 1945, challenging architects to design prototype houses using components donated from local manufacturers for reproduction on a massive scale. He was also determined to end speculation about the shape of low cost housing for the masses by ensuring that the houses were actually built, for genuine clients in real life conditions. Between 1945 and 1966 a total of 36 prototypes were designed. Not all were built, but even those that remained on the drawing board reveal how the architectural profession was increasingly willing to explore the potential of inexpensive, unpretentious, industrial building materials.

The first six were complete by 1950, but while their popularity was not in question – they drew thousands of visitors – none of the Case Study houses ever went into production as Entenza had hoped. Instead, the built survivors are revered as masterpieces of modern design, which was certainly not the intention.

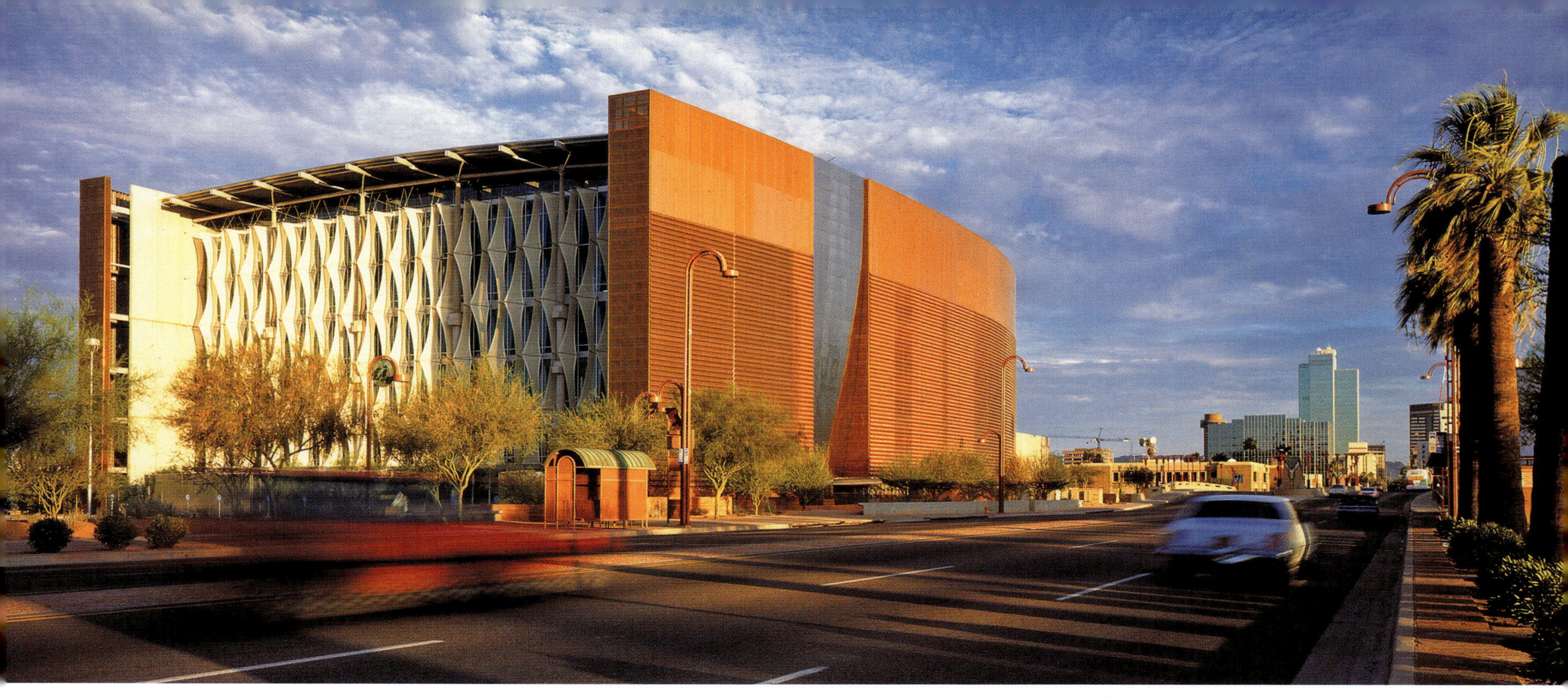

OPPOSITE PAGE AND ABOVE Phoenix Central Library by Will Bruder (1995). Corrugating copper distorted perceptions of a profile commonly associated with agriculture.

FOLLOWING PAGE Local farm buildings inspired the form and materials of the premises for Wild River, a company that offers white water raft trips. To 'activate the patina' of the corrugated steel, the building was sprayed with an ammonia solution.

CSH8, the Eames House in Pacific Palisades – self-designed as a home and studio by Ray and Charles Eames – is a case in point. The two-storey building was built with components selected from the Truscon Steel Company catalogue; its steel frame supported windows and wall panels and a roof of metal sheeting. The whole was lightweight, cheap and easy to build.

At the southern end was a full-height, open-plan living space, with bedrooms accessed via a spiral staircase. It was separated from the studio, at the other end of the house, by an enclosed courtyard. Following completion, in 1949, the designers were photographed at leisure, surrounded by their belongings, promoting the idea that the house was suited to the needs of a busy modern couple. But the message did not resonate with the masses; the house was never replicated.

Another landmark of the Case Study House program was completed in 1960 by Pierre Koenig, a Californian architect who had designed CSH21 two years before. The Stahl House (CSH22) is a minimal steel-framed pavilion perched on a promontory in West Hollywood with a roof of five-inch-deep profiled steel that frames panoramic views of Los Angeles.

CSH22 has become one of the icons of the CSH program, a status cemented by the photography of Julius Shulman. Soon after its completion, Shulman took an evening shot of two young women talking in the open plan living space, with the Los Angeles city grid in the background. It is has become one of the most famous architectural pictures ever taken.

In Shulman's picture, the elegant gravity-defying living space is sandwiched by a sun lounger at the bottom and the metal roof at the top. Barely a decade earlier these products would have sent out conflicting messages of destitution and indulgence. In the photograph the two are intertwined, a reminder of how fast perceptions of the material were changing.

Also working with corrugated sheet metal in 1960s California was Frank O. Gehry, who had relocated from Toronto after the Second World War and has now become one of the few architects to have attained the status of a household name. Sheets of metal twisted into highly expressive forms have become one of his trademarks, and his experiments with corrugated steel during the late-1960s were instrumental in helping him to define this signature style.

The O'Neill Haybarn (completed 1968), part of a masterplan for a ranch at San Juan Capistrano in California, was perhaps the earliest of these experiments. Gehry's roof of corrugated steel, tilted between corners posts (telephone poles) elevated this simple, low-budget structure to the status of sculpture. The idea of distorting the flat forms of sheet metal to reflect the shapes and colours of the local landscape was taken further in Gehry's design of a studio and residence for artist Ron Davis, in Malibu (completed in 1972). The trapezoidal roof slopes from 30 feet to 10 feet in the opposite corner to help frame the mountain and ocean views.

While the West Coast of America retains strong connections with corrugated metal, the material also resonates with contemporary architects working in Texas, and the south-western states, notably Arizona. In these arid, frontier areas, where construction materials were once hard to come by, corrugated sheet metal has long been a prominent feature of the landscape. In many small communities, corrugated iron is as old as anything in the landscape, a point harnessed by architects eager to shape local identities and develop a regional vernacular architectural language.

Will Bruder (born 1946), an architect with a background in engineering and sculpture, has exerted a particularly strong influence on the use of corrugated metals in Arizona, both in

CAUTION
HIGH VOLTAGE

ABOVE AND RIGHT The 'Xeros House' on the outskirts of Phoenix, Arizona (2005). Although a common feature of America's southern states, the use of corrugated metals in residential architecture still retains the capacity to surprise.

residential projects and large-scale public schemes. 'Corrugated metal has always been a feature of my work. I used it in my first building, an addition to my own house in 1974, and I'm still using it today. I like giving ordinary materials new meanings.' Bruder has also exploited the symbolic significance of corrugated sheet metal. At Phoenix Central Library (completed in 1995), arguably his best known building, he was challenged to build, 'a monumental-scale building in the centre of the city on a very low budget [around US$100 per square foot]'. I wondered how I could clad this large container for books in a way that was appropriate for the setting, but also used a technology that was spare and repetitious, making it affordable. My mind turned to grain silos.' But instead of using corrugated steel, Bruder approached Bayliss, a manufacturer of grain silos based in Nebraska, with the idea of using their rolling machines to corrugate sheets of copper – Arizona used to be known as the 'Copper State'. The idea worked well. So sheet copper was ordered from Eastern Europe, sent to Nebraska to be rolled and then attached to the side of the library as a rain screen. Bruder describes it as, 'a metaphor for the landscape'.

The Library became one of the most published buildings of the 1990s, and is a well-established local landmark, 'but if I had used corrugated steel, the response would have been very different'. The decision to dress an expensive metal in industrial clothes distorted perceptions of the material; he created something new out of something commonplace.

Bruder is one of the dominant architects in Arizona and his influence has been substantial. Many of his former colleagues and employees have worked with corrugated metal, including Rick Joy, the firm of Wendell Burnette and Matthew Trzebiatowski of Blank Spaces, designer of the 'Xeros House' (Greek for 'dry') on the outskirts of Phoenix, which is sheathed in rusting corrugated steel and wire mesh. The steel was selected because:

> It was cheaper than other options, it is also very strong and looks beautiful in the landscape, although there has been some opposition to it. A lot of people associate rust with decay; it can be difficult to accept that the material is just protecting itself.

TOP AND ABOVE Urs Peter Flueckiger's decision to sheathe his West Texan home in corrugated steel was inspired by agro-industrial buildings in the local landscape, notably cotton gins

Urs Peter Flueckiger, a Swiss architect, has also encountered bemusement for his use of the material. Flueckiger, who settled in Lubbock, the heart of the Texan cotton fields, in the late-1990s, decided to build his home of corrugated sheet steel in homage to the large corrugated steel cotton gins that are such a prominent feature of the landscape but he found that local people associated the material with industry.

As in most parts of the world, ties between corrugated metal and American agricultural communities are deeply ingrained. Corrugated iron sheds rusting gently into the landscape are a familiar feature of many country areas. In recent years a number of contemporary architects have played on these associations for a variety of reasons, and with a broad range of outcomes.

For much of his career Samuel Mockbee (1944–2001) was committed to applying his skills as an architect to the design of low-cost buildings for families and communities stuck in a cycle of poverty, particularly in Alabama. He minimized costs by using volunteer labour – often the client community – and building with salvaged or donated materials. Corrugated steel, much of it recycled, featured regularly in his buildings, many of which were revered as much for their architectural quality as for the ethical spirit that underpinned them; Mockbee was posthumously awarded the American Institute of Architects' Gold Medal in 2003.

ABOVE AND BELOW When not in use the sliding doors of the Zachary House in Louisiana can be closed so that it resembles an agricultural shed.

In Louisiana, architect Stephen Atkinson has used the material for quite different reasons: as a discreet weekend house for his parents near the small Louisiana town of Zachary. The Zachary House sits on a long rectangular timber base, bisected by a deck projecting at a right angle. Corrugated sheets of untreated steel and fibreglass are attached to the walls and roof of its timber frame. When in use, walls slide back to reveal the entrance and windows; at other times, the building can be closed up to resemble a barn. The only other mark that the Zachary House makes on the landscape is a freestanding red brick 18-foot chimney.

Australia – the expression of a national spirit

Australia is, beyond any doubt, the spiritual home of corrugated iron. There are countries that produce more of the material, there may also be states in which corrugated metal is a comparably prominent feature of the landscape, but the factor that sets Australia apart is the predominantly positive light in which it is perceived.

The development of Australia since European settlement and the evolution of corrugated sheet metal as a construction material run almost side-by-side. The 'First Fleet' arrival at Botany Bay in 1788 predated the invention of the material (1829) by only

Woolshed at Lake Mungo, New South Wales. Australia is the world's spiritual home of corrugated iron. The material has been a feature of almost every settled rural and urban landscape since the 1830s.

41 years. Within a decade, Richard Walker (see Chapter 1) had exported a number of corrugated iron storage buildings to the South Australian Company. Australia's appetite for the material has been unsated ever since.

Throughout the nineteenth century British suppliers of corrugated iron sheeting and prefabricated buildings thrived on exports to farmers, miners, sheepshearers and other settlers in the Australian colonies. But early attempts to establish an iron making industry in Australia failed. Poor quality raw materials, high fuel costs and unpredictable markets were among the many limitations for local producers. It was not until 1914 that sheet steel was successfully produced on Australian soil – in Newcastle, New South Wales. Today, Australia is one of the world's leading producers of corrugated steel products, a status reflected in almost every settled urban and rural landscape across the country.

Corrugated metal is used in water tanks, sheep pens, fences, all manner of roofs and sheds, even a honeymoon hideaway. The 'Honeymoon Hut', in an isolated corner of inland South Australia, dates to the 1880s, when it was built as accommodation for riders patrolling the state boundary. Legend has it that local stockmen would take their wives to the rough and ready shelter on their wedding nights.

A measure of corrugated iron's significance to the national sense of self came during the opening ceremony of the Sydney Olympic Games of 2000, an opportunity for the host nation to show its best side to the world, which included dancing girls holding sheets of corrugated metal as props in a spectacle titled 'Tin Symphony'. But favourable associations between Australia and corrugated iron have only been articulated relatively recently.

The 1960s was a tumultuous decade for many countries, and Australia was no exception. The government's support for the American invasion of Vietnam had caused divisions at home and led to scrutiny of Australia's allegiances on the world stage. The 'cultural cringe', a term coined in 1950 by Melbourne critic and writer A. A. Philips to describe the Australian tendency to regard local culture as inferior to work produced overseas, was another cause of self-analysis. Social issues were also a source of consternation, notably the treatment of Australia's indigenous population. These and other factors created a collective conundrum, namely, what does it mean to be Australian? With the benefit of hindsight it seems that corrugated iron had a role to play in addressing this question.

The material had been a constant feature of the landscape for over a century. It has even been claimed that corrugated iron helped to shape the landscape: without corrugated iron water tanks, which created portable, independent water sources, settlement of the arid interior would have been far more restricted. More pertinently at this time of soul-searching, corrugated iron seemed to embody values at the core of Australia's nascent national identity, including honesty, practicality, strength and independence. In 1981 the point was taken further by the architect Peter Myers in a polemic published in the cultural monthly *Transition*:

> The culture of this country has [been], and will continue to be moulded by corrugated galvanised iron... We have imbued this material, with its consistent capacity to fabricate simple solutions to complex problems, with a degree of omnipotence which is entirely consistent with its claim to primacy in our still infant culture.

There were at least two strands to the revival of interest in corrugated iron. One was related to architects' attempts to create a more empathetic relationship between European settlers and Aboriginal communities – an ambition actively promoted by the progressive Labour government led by Gough Whitlam (1972–75). Up to the early 1970s there had been few successful attempts to create appropriate shelter and meeting places for Aborigines. Things began to improve when the Europeans started to build their understanding of Aboriginal cultures and settlement patterns. One of the earliest collaborations between a European architect and an Aboriginal tribe began in 1973, when the Aboriginal Arts Board of Australia sent architect Peter Myers to Bathurst Island, about 50 miles off the coast of the Northern Territory. Myers was employed to work with the Tiwi people on the development of a 'keeping place' for the community's cultural artefacts.

The original intention had been to use traditional materials, but following Cyclone Tracy – which hit Australia's 'Top End' on Christmas Day 1974 – new building codes were introduced, which demanded the use of a steel frame. The outcome was a long open-plan volume lined in plywood, with a roof of curving corrugated steel described by author and academic Philip Good as, 'a modern echo of the sheet bark structures that the Tiwi people erected annually for Kulama, the ceremony to celebrate the ripening of yams'.

Among the theories about Australian Aborigines' acceptance of corrugated iron is that the material is a 'man-made bark', directly replaceable with bark sheets used in traditional dwellings.

There are many theories about the Aborigines' acceptance of corrugated sheet metal, including the idea that it is 'man-made bark'. In 1985 Australian architectural historian, author and teacher Philip Drew wrote:

> Aborigines prefer corrugated iron above all other materials for building... in their traditional shelters the Aborigines used bark in large sheets which were bent or shaped... to form shelter... it was therefore natural for Aborigines to adopt corrugated iron as an improved type of bark. (*Leaves of Iron*)

It was also natural for European architects to learn lessons from Aboriginal construction traditions. From the mid-1970s two principles in particular – a commitment to building in harmony with the landscape and an ambition for buildings 'to touch-this-earth-lightly', a saying of Aborigines in Western Australia – underpinned the architectural ethos of Glenn Murcutt, who is widely regarded as the originator of a contemporary Australian vernacular tradition. For many his name is also synonymous with corrugated iron (see also pages 180–91).

From the mid-1970s Murcutt began to use galvanized corrugated steel with meticulous care and precision in the design of discreet and finely detailed residential buildings. Fusing pure rationale modernism with an appreciation of the landscape, he cultivated a distinctively Australian style of architecture, which became the catalyst for another strand in the revival of corrugated sheet metal in the Australian context.

Murcutt's house for Marie Short at Kempsey in rural New South Wales, completed in 1975, was one of the first buildings to demonstrate his passion for site-specific modernism. The house is composed of two parallel barn-like pavilions (a reference to local agricultural buildings) sitting on timber stumps (to minimise the impact on the land) with overhanging corrugated steel roofs (another reference to local vernacular and helping to control the internal temperature) oriented to harness cooling breezes (to avoid the need for air conditioning). It was a seminal work (see pages 182–4).

Since the late-1970s, countless architects have used corrugated sheet metal all over the vast Australian landmass, with its broad variety of climates and landscapes. John Andrews used it to design a prototype prefababricated farmhouse for arid rural areas, reviving a tradition dating back to the 1850s. The building – which still stands in Eugowra, 220 miles from Sydney – is defined by a steel tower protruding periscope-like from the roof. The tower includes a lightning conductor, chimney and water tank – a familiar feature of the Australian landscape. It also incorporates provision for solar panels and a wind-powered motor to raise water from the ground-level water tanks. The farmhouse never went into production.

Corrugated iron is also a common feature of architecture in Australia's tropical north. In the years following Cyclone Tracy – which destroyed over 50 per cent of Darwin and coincided with a downturn in the economy – Australia's tropical regions witnessed a building boom. Adrian Welke and Phil Harris – who set up in practice as Troppo Architects – were among the architects who won work in the Northern Territories and Queensland during the 1970s and 1980s. Like Murcutt and Leplastrier before them, Troppo Architects, and many of their peers, including Russell Hall, Rex Addison, Lindsay Clare and Gabriel Poole, were committed to the development of regionally specific forms of architecture, which often involved re-learning forgotten skills. In the context of

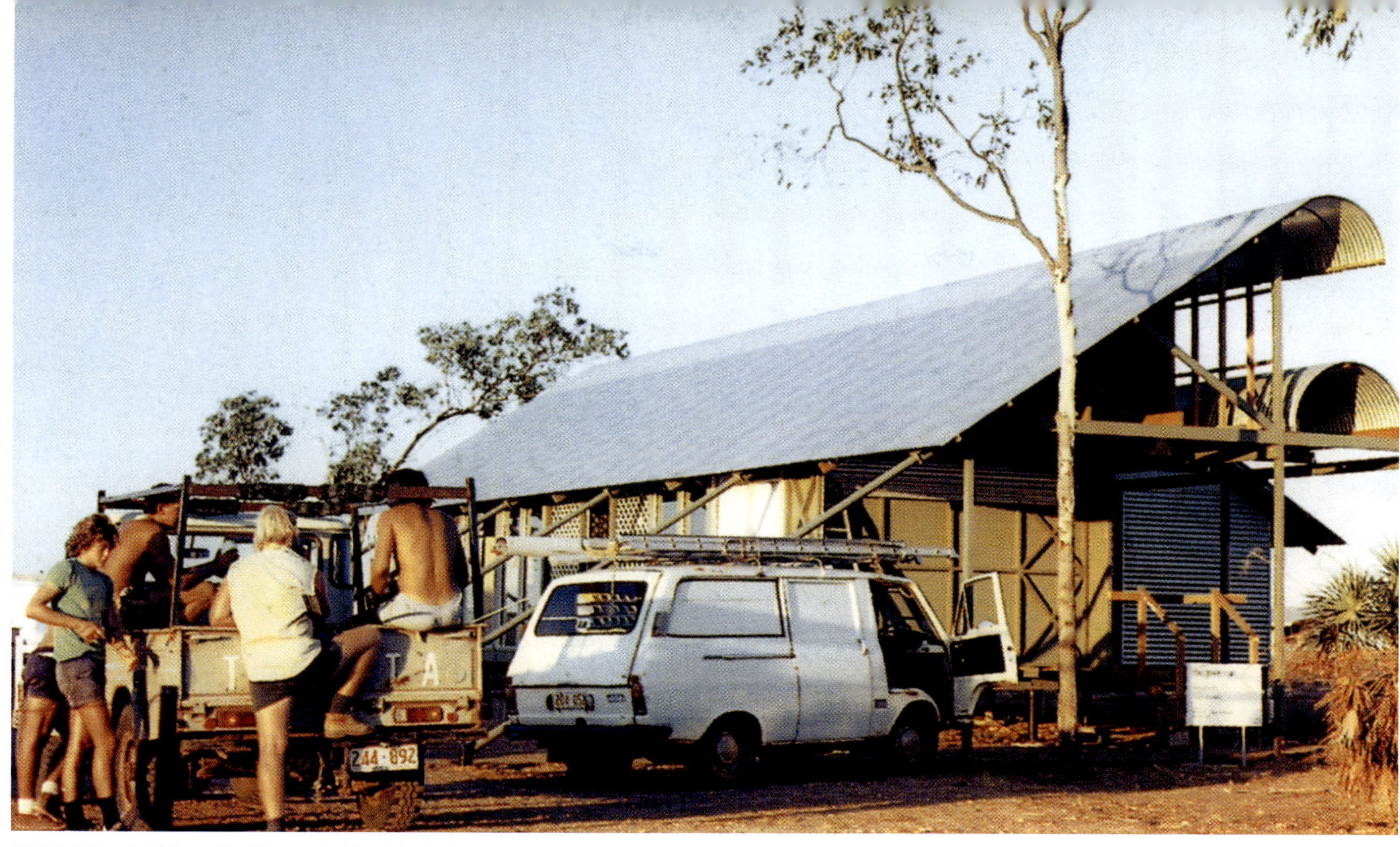

ABOVE AND RIGHT 'Green Can', Troppo Architects' entry in a low-cost housing competition in Darwin, 1981-82. Corrugated iron allowed the architects to minimize costs and shape the roof to catch cooling breezes.

BELOW The Kaiplinger House, 1983, one of a number of elevated homes sheathed in corrugated steel built by Troppo Architects in the Coconut Grove district of Darwin.

Darwin, which is closer to the equator than Bangkok, these included ideas such as raising houses on stilts to aid ventilation; designing adjustable skins (for instance, sliding doors, battens and louvers); avoiding roof linings, which can become homes for unwelcome creatures; orienting buildings to harness cooling breezes; and using easily transportable materials.

One of Troppo's first designs was the 'Green Can', an entry in a low cost housing competition in Darwin – the name refers to Victoria Bitter, an Australian beer that comes in a green tin. Its roof was defined by two sloping sections of corrugated iron at slightly different heights: the upper level contained the bedrooms and living spaces; the lower section contained the main services. A corridor between the two drew cooling breezes into the building. During the 1980s a number of houses based on the Green Can model were built in the Coconut Grove district of Darwin, which became known as 'Troppoville' – or 'shantyville' to less enthusiastic locals.

Corrugated iron was fundamental to the scheme in two respects: it helped to keep the house within the A$34,000 budget, and the flexibility of the sheets allowed the architects to shape it to the optimum angle to catch the winds (35 degrees). The material has been a consistent feature of Troppo's regionally specific architecture ever since. Aside from the material's affordability and flexibility, the architects have found it easy to

PREVIOUS PAGE The Rozak House, 2001, by Troppo Architects, occupies an isolated rocky outcrop an hour's drive south-east of Darwin. Its corrugated iron roofs have been 'twisted' to deflect strong winds.

BELOW Aerial view of Bowali Visitors Information Centre in Kakadu National Park, designed by Troppo Architects in association with Glenn Murcutt.

BOTTOM Glass-blowing studio at Nungurner, Victoria, designed by architect Don McQualter in 1999.

OPPOSITE AND FOLLOWING PAGE The Wheatsheaf House, near Kyneton, Victoria (2004). Architect Jesse Judd's use of corrugated steel bent into the shape of a wave, or bull's nose, refers to an Australian vernacular tradition evident in verandas and shop fronts across the country since the early nineteenth century. The interior is lined with red plywood, a stark contrast to the silvery messmate plantation that surrounds the house.

transport and to use within a kit of parts; age-old virtues of the material that remain of considerable value in isolated areas of the 'Top End'.

In 1994 Troppo joined forces with Glenn Murcutt on the Bowali Visitor Center, at Kakadu National Park, 155 miles east of Darwin. The team, which also included representatives of the Aboriginal community, reorganized, extended and partially demolished the existing structures and linked them with a long, low corridor deck, raised slightly off the ground and enclosed under a corrugated steel shell. The walls of the new buildings were made of rammed earth. The composition won great praise. In his 2005 monograph about the practice Philip Goad wrote: 'The long horizontal of the gable roof of silver corrugated iron provides an heroic view of the timeless rural shed, and in a certain light it becomes simply a brilliant reflective sky.'

Since then there has been no let up in the use of corrugated steel in Australia and it seems that no building type is exempt. Recent examples include an archery pavilion for the Sydney Olympics in 2000, a glassblowing workshop in rural Victoria and a private house in a secluded messmate wood (*Eucalyptus oblique*), as designed by Melbourne architect Jesse Judd in 2004. So perhaps the widespread assumption among Australians that corrugated iron is a local invention is hardly surprising.

Europe – corrugated iron returns to its origins

In many respects the reinterpretation of corrugated sheet metal as a feature of contemporary architecture in Europe followed the pattern established in post-war America, when young architects infused by the values of modernism looked to industry for answers to design challenges.

Like the Americans of the 1940s and 1950s, the generation of European architects that emerged during the 1960s was interested in the potential of prefabrication and interchangeable machine-produced components to assist in the development of new forms of architecture. British architects led the way. Notable exponents of the style that came to be known as 'high tech' included Norman Foster, Richard Rogers, Nicholas Grimshaw and Michael Hopkins.

There were home-grown influences on these architects, including Peter and Alison Smithson's kit-of-parts 'House of the Future', exhibited at the Daily Mail Ideal Home Exhibition of 1956, and the popular culture-inspired 'plug-in' architecture of Archigram, an influential collective of British architects fascinated with the possibilities presented by technology. On the other side of the English Channel, Jean Prouvé's innovative work with pressed metal sheeting and industrial components was an enduring influence, but arguably the primary inspirations were American, notably Buckminster Fuller, with whom Foster collaborated from 1968 until the early-1980s.

ABOVE AND LEFT Team 4's factory for Reliance Controls, Swindon. Lightweight, industrial components were used to create an early landmark of high-tech architecture.

While completing post-graduate studies at Yale University in the 1960s, Foster and Rogers became close friends. During their time in America they travelled to South California – for a brief spell Rogers worked in San Francisco – where they explored the architecture of Craig Ellwood, Charles and Ray Eames, Pierre Koening et al. These spare steel structures, with their exposed frames and engineering-led discipline, left a lasting impression.

When they returned to Britain, Foster and Rogers set up in practice as Team 4, originally with sisters Wendy and Georgie Cheesman, also Yale alumni. Perhaps the most revered outcome of their short-lived collaboration, and the purest encapsulation of their mutual interests in industrial components and prefabrication, was a factory for electronics company Reliance Controls in Wiltshire, on which design work began in 1964. The 30,140 square feet (2,800sq m) flexible space was made of lightweight industrial components, and was clad entirely in profiled metal sheeting. It was an early landmark of 'high tech' architecture. Team 4 disbanded in 1967, but profiled metal remained a prominent feature of both Rogers' and Foster's output for years to come. In 1968 Foster began work on two low-cost buildings for Fred Olsen, a Norwegian cruising and freight company at Millwall Dock, a depressed area of East London. A monograph on Foster's work describes the buildings as an, 'appropriately machine-like presence in a wholly industrialized landscape'.

ABOVE Roof of the Fred Olsen Amenity Centre, Millwall Dock, designed by Foster Associates, 'an appropriately machine-like presence in the wholly industrialised landscape'.

RIGHT The Passenger Terminal for Fred Olsen was composed of two steel-framed tubes clad in corrugated aluminium.

The Amenity Centre, a processing centre for goods and passengers, was notable for its glass curtain wall; the Passenger Terminal, reminiscent of the Airstream trailer, was a simple structure composed of two long steel-framed tubes, wrapped around two sides of the Amenity Centre. Early studies of the Passenger Terminal – rejected for reasons of cost – envisaged a tube made of two skins of cold-rolled plastic-coated profiled steel decking, similar to that used at Reliance Controls. Instead, a much lighter structural system was used, in which the tubes were clad in a single skin of corrugated aluminium.

At around the same time, on the other side of the London, Nicholas Grimshaw specified corrugated aluminium sheeting on an altogether different type of building in a far more prestigious setting: a low cost apartment block funded by a housing association overlooking Regent's Park. The Park Road Apartments tower was the first residential building in Britain built with a central core, which freed up the interiors and simplified the design of the perimeter glazing – sourced from the company that supplied the glass used on London buses. The balance of the envelope was made of high-grade aluminium sheeting, with distinctively curved corners. The solution was lightweight, cost-effective and appeared progressive. In a 1995 monograph about Nicholas Grimshaw and Partners, Colin Amery

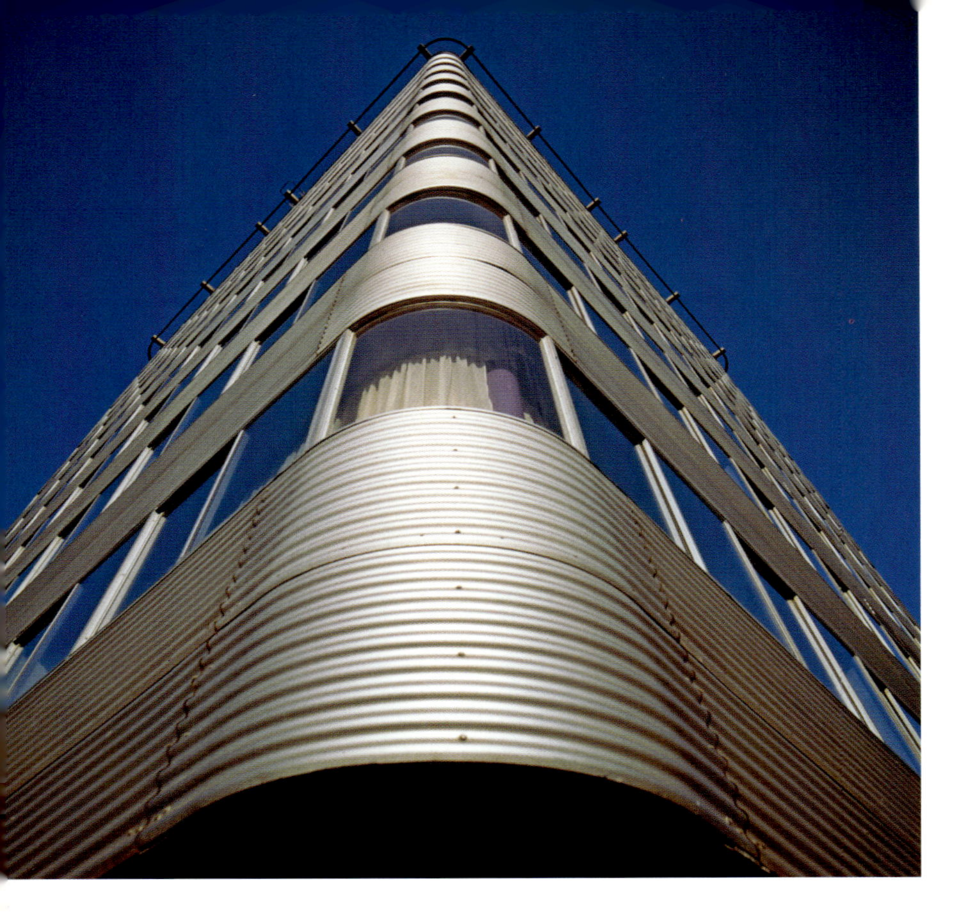

Park Road Apartments, Nicholas Grimshaw, *c.*1968: The building's central core allowed the use of a repetitive envelope of glazing and high-grade aluminium sheeting.

explained that, '. . . at the time. . . sinusoidal cladding. . . was considered adventurous because it demonstrated certain advantages of an architectural "lightness of being".' It also reflected the avant-garde spirit of its predominantly youthful and creative occupants – designers, photographers, architects – all of whom shared in the ownership of the building, including Nicholas Grimshaw himself, who lived there for six years with his young family.

Beyond the rarefied confines of the metropolitan cultural elite, corrugated sheet metal carried quite different messages. For many people it remained synonymous with poverty or agriculture. These negative connotations were partly responsible for a rare low point in Norman Foster's illustrious career. In the early 1970s Foster Associates was appointed to design a housing development on a 100-acre site at Woughton, south of Milton Keynes' city centre. The Midlands conurbation had been designated a new town in 1967. Its subsequent development was based on urban design principles imported from America; the predominantly low-rise car-inspired garden city was built on a most un-British grid.

In early 1970s housing was the priority for Milton Keynes; 3,000 new homes were required each year. As a result labour was in short supply; funds were also limited. The stage was set for the use of a prefabricated housing system.

Foster researched agricultural and industrial system-built models before cost constraints influenced the use of the 'Quick Build' timber frame system, devised by Walter Llewellyn, which allowed shells to be erected and roofed in only one day. The external walls had a plywood skin, lined with bitumen-impregnated insulation and clad with profiled aluminium sheet. The use of low-pitched aluminium roofs was another issue driven by cost limitations rather than the architect's wishes – preliminary designs included mono-pitched roofs.

Phase 1 of Bean Hill was completed in 1973. It consisted of 492 single-storey and 72 two-storey houses grouped in terraces around cul-de-sacs. But it was not a success. A tenure ratio of 75:25 in favour of homes for social rent meant that Bean Hill attracted a high proportion of disadvantaged families, and these people, at the lower end of the socio-economic scale, were sensitive to the appearance of the corrugated aluminium, which, for some, was reminiscent of post-war temporary housing. But the legacy of corrugated iron's perception was only one factor that undermined the success of Bean Hill; the houses also performed poorly.

Corrugated aluminium featured prominently at Bean Hill, a large housing development in Milton Keynes, *c.*1973, but the material was not popular with its predominantly low-income inhabitants.

LEFT Detail of the original external walls at Bean Hill, Milton Keynes.

ABOVE AND RIGHT The Vitra Furniture Factory designed by Nicholas Grimshaw in 1981 was ready for production only six months after its predecessor had been destroyed in a fire.

Today, Bean Hill's identity and appearance have been comprehensively overhauled. The predominantly owner-occupied development is known as 'The Gables', the houses have been re-roofed with tiles and slate and the aluminium cladding has been removed.

While the Bean Hill episode was a reminder that the use of corrugated sheet metal still demanded sensitive handling, particularly in the design of social housing, buildings for commerce and industry were another matter. In 1981, only six months after a fire had destroyed the Vitra Furniture Factory at Weil-am-Rhein in Germany, Nicholas Grimshaw had completed a new facility; the large factory building was sheathed entirely in profiled aluminium. The speed and efficiency of construction harked back to the experience of the early Victorians. Grimshaw also used the material in a headquarters building for German industrial design company MABEG, in Soest. Recalling the architect's Park Road Apartment tower, the building has curved corners and is wrapped in an envelope of corrugated aluminium, a response to the client's brief for a building that was, 'functional, flexible and visually striking'. Corrugated metal was also used by fashionable French designer/architect Philippe Starck, in the design of a factory for Forge de Laguiole, a knife and cutlery manufacturer in the small town of Laguiole, in the Aveyron region of south central France. In its original format, a 60-foot steel blade rose from the flat roof of the simple, metal structure.

By the early 1990s the rehabilitation of corrugated sheet metal in Europe was all but complete. To leading edge architects the material, in all its various profiles, colours and textures, communicated any number of messages, including urban chic and environmental sustainability.

PREVIOUS PAGE Nicholas Grimshaw's headquarters office for MABEG in Soest, Germany *c.* 1999, whose curved envelope of glass and aluminium recalls the Park Road Apartments tower of 1968.

ABOVE AND RIGHT Rem Koolhaas used lacquered corrugated aluminium to define two apartments within the Villa Dall'Ava in Paris.

FOLLOWING PAGE The perception of corrugated metal as a temporary or 'nomadic' material was harnessed by Foreign Office Architects in the Bluemoon Aparthotel in Groningen in 2001.

In 1991 the fashionable Dutch architect Rem Koolhaas used sheets of corrugated aluminium lacquered with copper and aluminium to define two separate apartments within the Villa Dall'Ava, a luxurious and structurally challenging house in the Saint-Cloud district of Paris. A decade later, in a dense corner of North London, architects Sarah Wigglesworth and Jeremy Till used corrugated steel and plastic sheeting to shield straw insulation from the elements in the 'Straw Bale House', an example of low-impact residential architecture.

At around the same time, in the Dutch town of Groningen, Foreign Office Architects, a highly regarded London-based practice, harnessed the age-old perception of corrugated metal as a transient, impermanent material in the design of a small hotel in a square in the town's old port district. The four-storey structure, which originally comprised three suites and an entrance lobby, is clad entirely in sheets of corrugated metal, each of them on hinges. Like a tent, the building is intended to telegraph its level of occupancy: when full, the corrugated shutters are opened and the hotel glows in the dark; when the hotel is empty, the building is 'zipped up,' and takes on the form of a mute box.

The playful symbolism and delicate handling of corrugated sheet metal at the Bluemoon Aparthotel in Groningen could hardly have been more different from Richard Walker and Henry Palmer's use of corrugated iron at the London Docks in the 1830s. By the beginning of the twenty-first century, perceptions and uses of the material had come a very long way from its origins as a structurally innovative roofing material.

GLENN MURCUTT

Australia

Farmhouse for Neville Fredericks, Jamberoo, New South Wales (1982). Glenn Murcutt's work has become synonymous with a commitment to developing forms of contemporary architecture appropriate to Australia's landscape, climate and lifestyles.

> It is very easy to misunderstand the importance of corrugated iron to [Glenn] Murcutt's architecture. It is not a gimmick or a cheap trick to attract notice. Rather it arises from his sense of the beauty and poetry in ordinary things and his desire to create buildings that speak to the people. His rediscovery of corrugated iron… [was] not an end in itself… [but] a sensible response to [something that had] been forgotten… and which… could not have been bettered. (Philip Drew, *Leaves of Iron, Glenn Murcutt: Pioneer of Australian Architectural Form*, 1985.)

No architect is as closely associated with corrugated iron as Glenn Murcutt (born 1936). The material has been a consistent feature of the Sydney-based architect's idiosyncratic and uniquely Australian work since the mid-1970s, the point at which he began to develop his distinctive architectural voice.

As well as appreciating corrugated iron's practical qualities and the emotional bond between the material and Australian vernacular buildings, Murcutt's interest in corrugated iron has been driven by a commitment to developing forms of contemporary architecture appropriate to the Australian climate and landscape. For him the material resonates both with native Australian fauna, which Murcutt has described as, 'durable, hardy and yet supremely delicate', and with native trees, '[whose] high oil content ... combined within the strong sunlight results in the foliage shimmering silver to weathered greys'. These adjectives, 'durable', 'hardy', 'shimmering', could equally be applied to galvanized corrugated metal.

In his early years, Murcutt, a seasoned traveller, also took inspiration from the pared-down purity of high modernism, the harmony between traditional Greek architecture and the landscape and Finnish vernacular architecture, particularly as interpreted by Alvar Aalto.

The first built synthesis of these diverse aspirations and influences was the Marie Short farmhouse, two long single-storey pavilions with corrugated metal roofs, completed in 1975 at Crescent Head near Kempsey, a town about halfway between Brisbane and Sydney. Philip Drew described it as, 'a mutation of Mies van der Rohe's… glass pavilion type, which Murcutt transformed by merging it with the primitive hut archetype.'

Marie Short's original brief was for an extension to the existing farmhouse, located at the base of a hill. Murcutt convinced her of the merits of a new building on a terrace at the top of the hill, which he regarded as a much better position. She agreed, but insisted that the house be demountable, in case there was a need to relocate it at a later date. Another condition was that the house be as cool in summer and as warm in winter as standing under the mulberry tree next to the old farmhouse.

ABOVE Murcutt's farmhouse for Marie Short, near Kempsey (1975), combines references to high modernism and Australian vernacular dwellings. It was the first of several private homes arranged around long, staggered pavilions with roofs of corrugated metal designed by Murcutt from the mid-1970s.

OPPOSITE Murcutt, pictured left, on site at the Marie Short farmhouse *c.*1975. In 1980 he bought and extended the building.

So Murcutt immersed himself in understanding the site and the immediate microclimate, establishing a pattern that has been replicated ever since – for him, 'the land matters more than anything'. At the Marie Short farmhouse this influenced the east-west orientation of the two staggered pavilions. It is also where the logic of the corrugated steel roof with its distinctive saddleback profile came in: the 39-degree pitch helps to deflect the cold winds and draws the warmer summer breezes into the house.

The use of timber and corrugated iron in a modern house was extraordinary; until the mid-1970s European building traditions generally held sway for the majority of cultivated Australians. Both materials have now become trademarks of an authentic 'Australian' style of architecture.

The building's performance was also a carefully considered response to local environmental conditions. Murcutt was determined not to rely on air conditioning; instead he proposed to harness the wind, rain and sun to create a liveable building in harmony with the elements. This explains the long sidewalls, which have been designed to facilitate cross-ventilation; the large west-facing veranda, intended as an extension to the living room; and a 3 inch-thick layer of roof insulation to moderate the internal temperature. In his 1985 monograph on Murcutt's early work – Philip Drew described the house as, 'an observation post for experiencing nature'.

When the Marie Short farmhouse came up for sale in 1980, Murcutt bought it. He has subsequently extended it and uses the house as a country retreat.

For the remainder of the 1970s, and on into the 1980s, Murcutt developed and refined his style. He built a number of private houses in rural areas, applying a formula similar to the Marie Short farmhouse – pairs of long, single-storey pavilions built with timber and corrugated metal – but always with distinct differences inspired by the landscape and climatic conditions.

The corrugated iron roof of Henric Nicholas' farmhouse (completed in 1980) in the shadow of Mount Irvine in the Blue Mountains, west of Sydney, has an angular roof profile, pitched at 45 degrees. The reasons for this are both practical – the steep pitch is better equipped to shed snow: and symbolic – the jagged profile echoes the silhouette of the nearby mountains.

Other corrugated iron features of the Nicholas farmhouse are six large water tanks; 'gal iron' tanks are a common feature of rural Australia, they enable isolated properties to be self-sufficient. And a curving rear wall, which shields the building from cold mountain winds. Murcutt has likened the wall to, 'an old man in an overcoat who has his back to a strong wind; he pulls the coat up around his neck for added protection but leaves it open at the front to admit the winter sunlight'.

One of the earliest sales points for corrugated iron was its incombustibility. In 1839 Richard Walker advertised corrugated iron structures with a 'double lining', claiming, 'the air space between being an effective barrier to the greatest intensity of flame' (see Chapter 1). Over 140 years later Murcutt completed his first all corrugated metal building, a house and gallery for two painters – Lyn Eastaway and Sydney Ball – in a bush fire-prone area of woodland near Sydney. The sheets of metal were not the only means of protection. Murcutt also installed external sprinklers to dampen the house when fire threatened.

The Ball-Eastaway house (*c.*1983) appears to be suspended in the landscape; the woodland is visible all around and the house makes only the slightest impression on the land. At the north-east end of the long, squat pavilion, where the site slopes away, the steel frame is supported on slim metal posts; at the other end it rests on a stone terrace.

Murcutt is best known as an architect of private houses, particularly country properties, and always in Australia – he has never built outside the country and has no intention of changing his mind. However, he has designed some public buildings. Two of his earliest include a restaurant at Terrey Hills, a suburb north of Sydney, and the Museum and Tourist Information Centre at South Kempsey. The latter has been described as his 'first major work'.

The original brief was for a museum; a small theatre and tourist office were added during the long design phase – work began in 1976

Marie Short farmhouse. Long sidewalls facilitate cross ventilation and timber stumps minimize the impact of the building on the site. The pitch of the roof deflects cold winter winds.

ABOVE The Fredericks farmhouse occupies the site of a former brick dwelling pulled down in the 1950s. Only the original fireplace remains.

OPPOSITE TOP AND BOTTOM The Henric Nicholas farmhouse, near Mount Irvine, New South Wales (1980). Murcutt has likened the curving rear wall to, 'an old man in an overcoat who has his back to a strong wind'. The front is open, to admit sunlight.

and the museum opened in 1982. Their seamless integration into the scheme demonstrates the flexibility of Murcutt's multi-pavilion methodology. In their original form, the three corrugated metal-sheathed pavilions enclosed a usable floor area of over 400 square metres (4,300 square feet). This was achieved while making the building appear incredibly light and undemanding in the landscape.

Other public buildings have followed, including the Bowali Visitor Centre, at Kakadu National Park, east of Darwin (*c*.1994), designed in collaboration with Troppo Architects (see pages 160–64). But the bulk of Murcutt's work remains residential.

He has an idiosyncratic approach to his work. He practises alone and selects his clients carefully: 'I like to meet the client informally and see whether we really belong together.' One consequence of this rigorous and intense design process is a rare degree of consistency. Murcutt's work is admired both in Australia, where he is widely respected as an international figure who is dedicated to his homeland, and around the world – he was awarded the prestigious Pritzker Prize for architecture in 2002.

Corrugated iron has only ever been one aspect of Murcutt's work, albeit a very visible one, but his 'rediscovery' of the material has made an immeasurable contribution to its re-evaluation as a component of fine contemporary architecture.

OPPOSITE Murcutt's house for painters Lyn Eastaway and Sydney Ball (1983) appears to be suspended in the landscape.

ABOVE Even the balcony is lined with corrugated sheet metal, as protection against bush fires.

ABOVE The Museum and Tourist Information Centre at South Kempsey (first phase completed in 1982) was Murcutt's first substantial public commission. The flexibility of his multi-pavilion building type was well suited to the diverse roles required of the building.

OPPOSITE The detailing of each pavilion differs according to its function and position.

LAKE|FLATO ARCHITECTS

Texas, United State of America

The steel frame for the three bays of the Carraro House was salvaged from an abandoned cement plant. From left: car park; library/bedroom; limestone-clad living space and full-height 'porch'.

Lake|Flato Architects' work is often described as 'humble', 'pragmatic', 'honest', 'strong' and 'no-nonsense' – words that could equally be applied to corrugated sheet metal, a material that has featured prominently in their diverse range of buildings since 1984.

David Lake and Ted Flato met while working for O'Neil Ford, a Texan architect who has been described as, 'ahead of his time in recognising the importance of local climate and regional culture'. Following Ford's death in 1982 they formed a partnership and built on their mentor's legacy. Like Glenn Murcutt in Australia (see pages 180–91), Lake|Flato Architects has since earned international recognition for designing distinguished but unpretentious buildings in harmony with local landforms, climatic conditions and terrain, which in their case means the flat, grassy scrublands of South Texas. Another similarity with Murcutt is the use of corrugated iron in the development of building forms that have come to define a regional vernacular architectural identity.

Corrugated iron became an established feature of the Texan landscape during the second half of the nineteenth century, when settlers began to move west in increasing numbers. The land was well suited to ranching, the practice of grazing cattle, sheep or horses on large areas of pastureland. Alongside barbed wire, corrugated iron was invaluable to the ranchers. The wire enabled them to control their livestock while the sheets of metal were a means of protecting the animals from the searing heat.These pioneer settlers have proved an enduring inspiration to Lake and Flato: 'We admire their practicality... using the materials at hand in a spare, simple way.' Like the ranchers' hand-crafted barriers and farm buildings, the architecture of Lake|Flato is frugal but effective.

Another rancher tactic adopted by the architects is the use of recycled materials. In 1990 Lake|Flato completed a house in Kyle, Texas that included materials salvaged from a decommissioned 1920s cement plant. The steel frame, stair rails and ventilators were bought for US $20,000 and were used to create a 6,500 square feet (600-square-metres) home at a total cost of US$200,000. The Carraro House comprises three steel-framed bays: an open-sided car park, a two-storey, corrugated metal-clad library and bedroom and a limestone living space and screened porch enclosed within full-height glass walls.

OPPOSITE AND BELOW Corrugated sheet metal, limestone and industrial salvage have become materials synonymous with Texan vernacular architecture.

The combination of industrial salvage, locally quarried limestone and corrugated metal have become a Lake|Flato trademark, as has the orientation of buildings to minimise the impact of the southwestern summer heat and provide ventilation from prevailing breezes.

For the Air Barns (completed 1998), two stables for polo ponies in San Saba, the architects welded together recycled oil pipes to create the frame and sheathed the buildings entirely in corrugated steel. The precision with which these rudimentary materials were handled created sheds of rare elegance and refinement, perfectly in tune with their environment. Lake|Flato's use of corrugated metal has done a great deal to alter perceptions of the material, not only in the design of residential buildings, but also in its migration to urban areas. Where previously corrugated sheet metal was identified almost exclusively with agriculture and industry, Lake|Flato has used it in the design of numerous public buildings, including two in their home city of San Antonio: the Great Northwest Library (1996) and the AT&T Center (completed 2003), an 18,500-seat sports stadium, whose expressed concrete frame is in-filled with inexpensive materials, including six different profiles of corrugated metal sheeting.

The box section sheeting used as a sun-shading device over a public concourse outside the stadium was inspired by a cattle feed lot in Bovina, Texas. David Lake was out walking one afternoon and, 'saw the sheets of corrugated iron strung up on wires, to protect cows from the sun... it is an example of a direct correlation between observing how materials work in the landscape, and application on the ground'.

Initially the architects experienced a certain amount of resistance to the use of corrugated metal on civic buildings in urban areas, but over the years, as David Lake says, attitudes have softened:

> Today, corrugated iron, along with limestone, has become accepted as part of the local palette of materials. Used in combination they create a sort of poetic dialogue; mass versus lightness.

ABOVE AND OPPOSITE Air Barns, San Saba. Recycled pipes and corrugated metal combine to create barns of rare refinement and elegance.

ABOVE AND OPPOSITE TOP AND BOTTOM The Great Northwest Library (1996) in Lake|Flato's home city of San Antonio is an example of corrugated sheet metal being applied on a public building in an urban context. Traditionally the material has been associated almost exclusively with rural landscapes, but since the early 1990s perceptions of the material have become increasingly positive.

Weight was also one of the determining factors in the use of arched corrugated metal roofs at the Birding Center, the gateway to a native habitat reserve in the Rio Grande Valley, Texas (completed 2004). David Lake explains:

> We needed to build low cost, low impact structures, so a structurally efficient roofing system was vital. We selected Quonset hut roofs, a common feature on local farms, and put them together to resemble three long barns. The form also refers to the Spanish tradition of building long masonry vaults.

The arches span 32 feet, and because they form both structure and building envelope, they use 48 per cent less steel by weight compared to traditional framing, an important consideration in minimizing disturbance to the delicate site and re-establishing the natural bird habitat.

There are many reasons why corrugated metal makes sense to Lake|Flato Architects, including its historic associations with Texas, its capacity to reflect as well as block out the sun, its minimal weight and its affordability – 'in recent years, the cost of construction has risen a lot, which may be one reason why more architects are using corrugated metal'. But Lake|Flato also take a longer-term view. 'We like the question O'Neil Ford used to ask. . . "What will your building look like as a ruin?"' The chances are that the ruins of a Lake|Flato building – perhaps including limestone, local timber and dust-soaked corrugated metal – will look timeless.

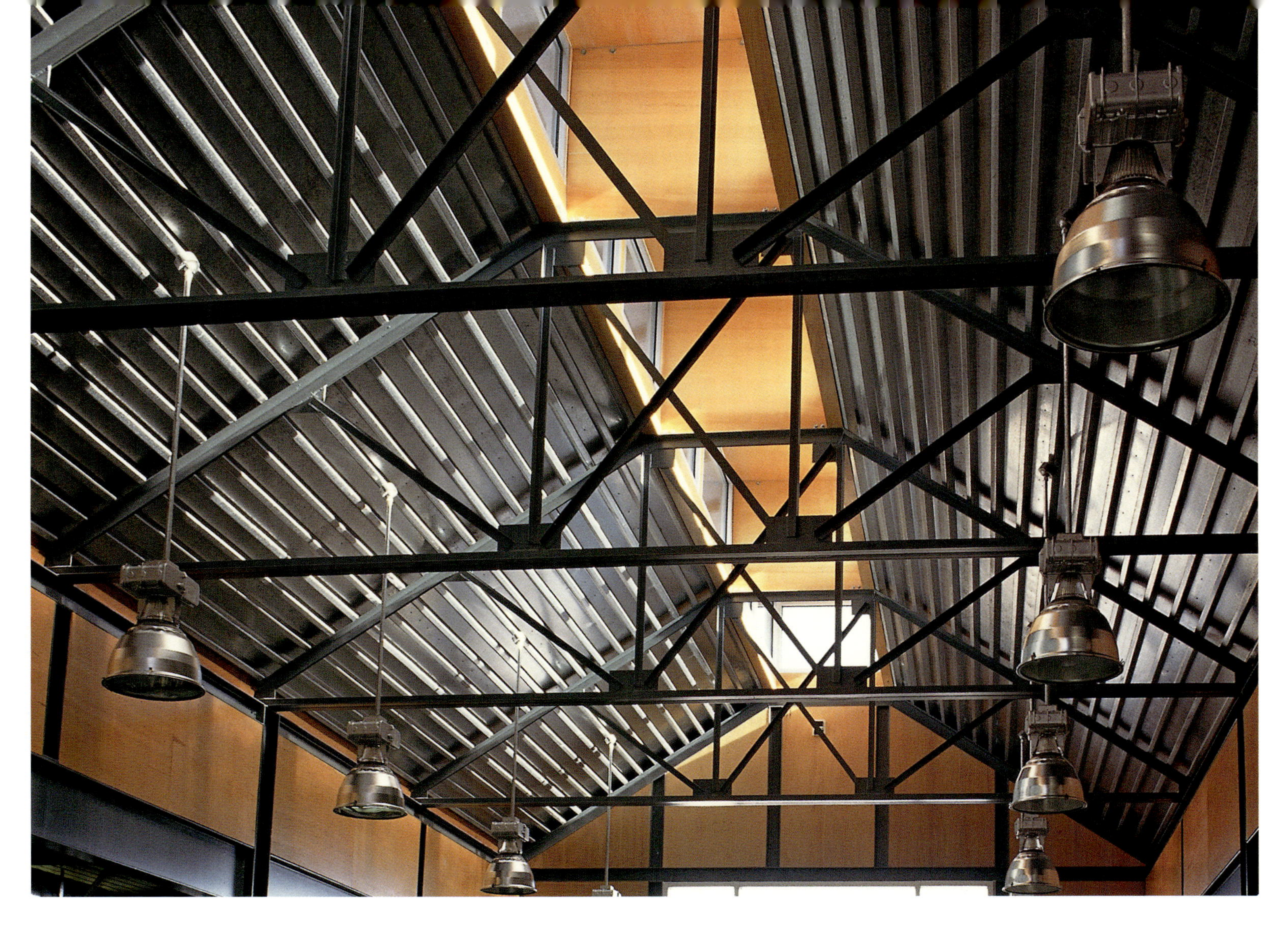

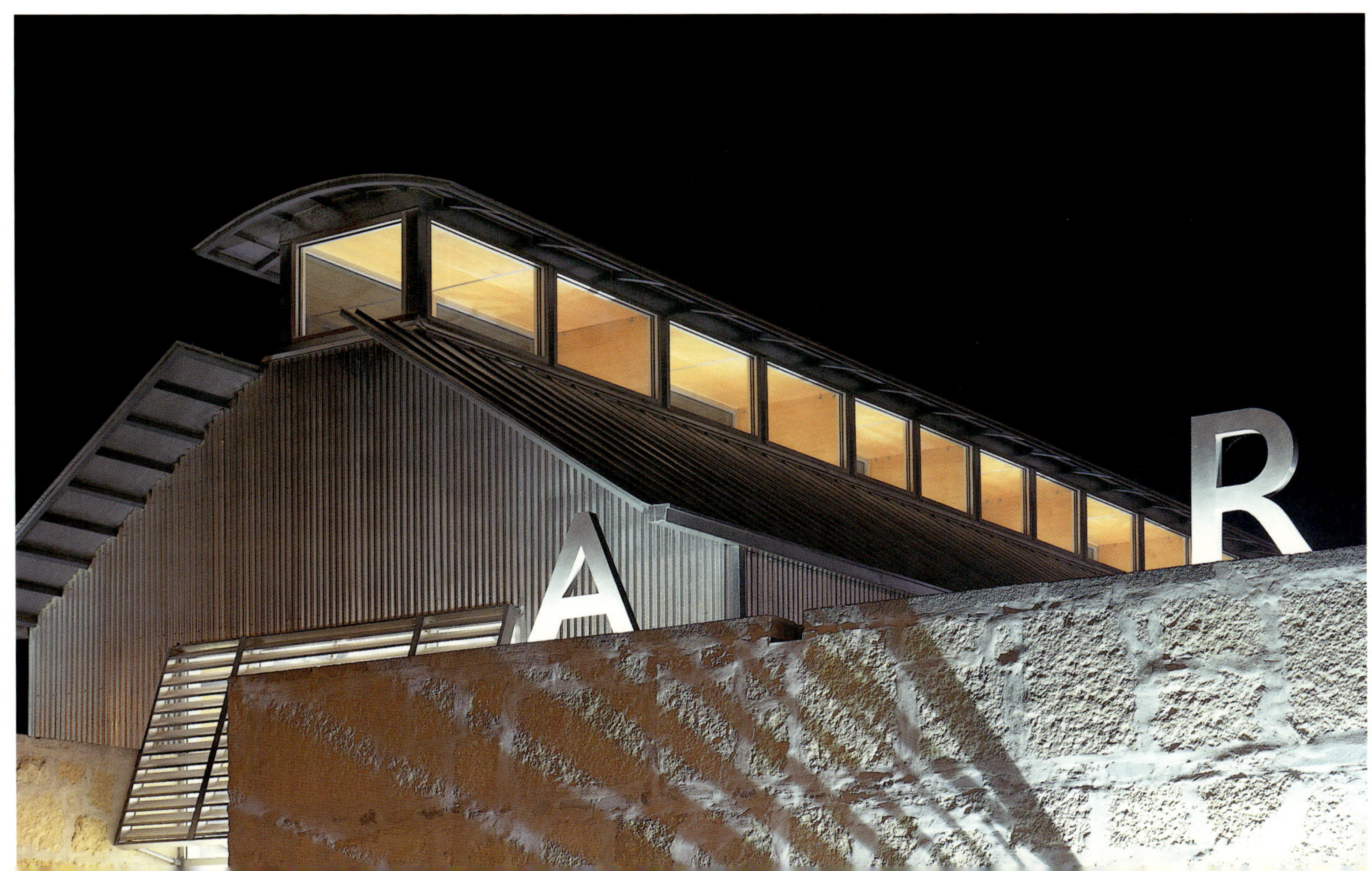
A
R

IT'S GAME

ABOVE AND PREVIOUS PAGE The box section sun-shading device over the public concourse at the AT&T Center in San Antonio was inspired by the use of corrugated sheeting at a cattle station at nearby Bovina (above). Six different metal profiles were used in the design of the stadium.

OPPOSITE TOP The Birding Centre is a lightweight gateway to a 1,700-acre native habitat reserve in the Rio Grande Valley. Flat sheets of steel shade visitors and vegetation from the sun.

BOTTOM The wide corrugations of the self-supporting roofs funnel rainwater into cisterns helping to irrigate the planted courtyards.

SHUHEI ENDO

Japan

OPPOSITE Sheets of corrugated steel bent into different shapes give an unmanned railway station outside Osaka an unexpectedly sculptural appearance.

BELOW Bike shelter at Maibara (1994). The rear convex wall is continued into the roof to create a wide, open volume, communicating the building's function and contents to passers-by.

Renmentai is a style of calligraphy common in Japan and China. It is a cursive script; the brush never leaves the paper. Entire texts, not just single words, are written as unbroken lines. Shuhei Endo (born 1960) uses *Renmentai* as a metaphor for his sinuous, free-flowing use of corrugated steel. 'My architecture... is based on... continuous surfaces or strips that form the outer shell, floors and roofing, in their continuity partly sharing, and shaping the complete buildings they define.'

In some respects this approach can be read as a contemporary interpretation of well established construction conventions. Endo explains: '[In] traditional Japanese architecture... spaces separated by sliding doors can merge to form a continuous space when the doors are removed... another way of defining a spatial hierarchy is to transform spaces by using folding screens.' Connections can also be drawn between the perception of corrugated metal buildings as impermanent and the Japanese tradition of periodic reconstruction. But in other respects Endo's use of corrugated metal represents something entirely new. Buildings of such expressive flamboyance are unusual anywhere; in Japan the use of the corrugated metal by architects has a very short history.

Corrugated metal panels were introduced to the country from the United States in the early twentieth century and they have been a staple

Sections through the house/clinic built for an Osaka pharmacist.

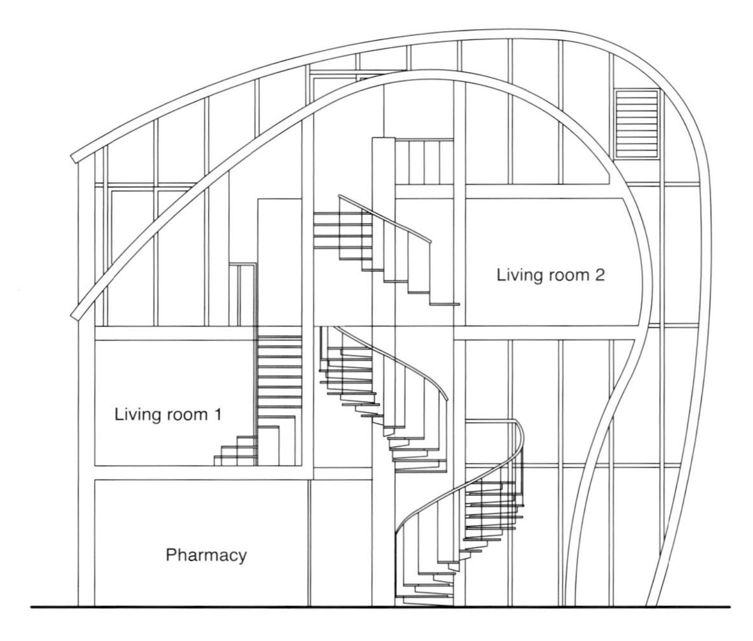

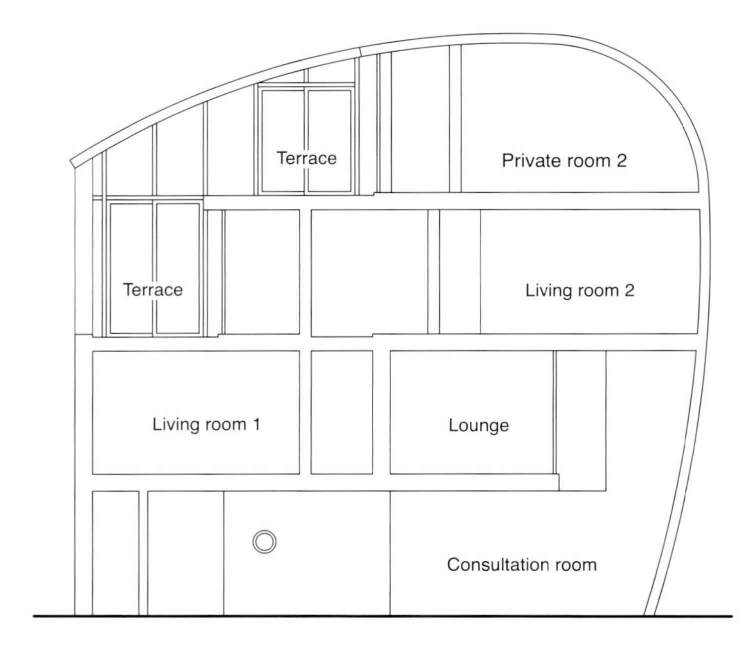

of civil engineering ever since. Growing up in the 1960s, Endo became aware of the material in irrigation canals, 'but only in the past 30 years has corrugated steel been used by architects'. One exception was Kenji Kawai, an engineer who in 1957 set himself the challenge of designing a home made only of recycled materials – he was working in a post-Second World War climate of resource shortages. The outcome was a large oval tube made of sheets of corrugated steel, which sat directly on the ground, without foundations or supporting columns. In 1975 architect Osama Ishiyama designed Gen-an, a villa inspired by Kawai's house. It was a tube with a corrugated steel envelope, but triangular in section, not oval.

Author, professor and engineer Hiroyuki Suzuki believes that in these houses,

> Kawai and Ishiyama established a tradition in Japan, showing how corrugated steel could be used in architecture. They converted a civil engineering material into a building material and opened up new possibilities in architecture.

Their legacy has been taken to another level by Endo, an architect fascinated by 'spatial hierarchy', 'permeability', 'continuity of space' and other theoretical concerns. He describes his buildings as, 'experiments in the use of corrugated sheet steel... [a material chosen] solely for technical and economic reasons'.

His first experiment, a bike shelter outside a railway station, was completed in 1994. The Mayor of Maibara, a small town about 40 minutes from Osaka, wanted a low-cost building that would accommodate 300 bikes and raise the quality of the urban landscape. Within this relatively open brief Endo was free to play with both materials and perceptions of bike shelters. The building is very simple: two levels of cycle parking and the attendant's office are enclosed within an unbroken arc of lapped and bent corrugated steel panels, which in profile resemble a question mark. Unlike typical bike shelters, where security is paramount, the Maibara bike shelter has only one wall; the building communicates its function and contents to passers-by, the intention being to encourage people to use it.

In another early project corrugated steel was used for its plasticity. An Osaka doctor wanted a house in a dense urban location that would include two apartments and ground floor accommodation for a pharmacy and a clinic for acupuncture and moxibustion – a form of heat therapy. Crucially, he also wanted the form of the building to reflect his belief in, 'the relatedness of all living beings... and the circulating equilibrium of bodily energies'. Essentially the three-storey structure was to be an advertisement for the doctor's practice. Endo's response was to design two three-storey volumes, staggered one behind the other and wrapped in a curvaceous envelope of corrugated steel. A full height glass facade projects a message of welcome and openness.

Endo has developed a code for naming his projects. For instance, 'Halftecture F', an unmanned railway station, refers to his interest in 'constructions consisting mainly of open space [half architecture]'; 'F' is the first letter of the location, Fukui. The design is extremely simple: sheets of corrugated steel bent into teardrop-shaped loops mark the ends of the long thin platform. Once again the material was used sculpturally, to create maximum effect with a limited budget.

The curvaceous envelope of corrugated metal and the open, circular form of the house/clinic reflect the pharmacist's beliefs in the, 'relatedness of all living beings'.

Shuhei Endo bent sheets of corrugated iron into three different profiles to create an undulating, ripple effect in the long, unmanned station near Osaka.

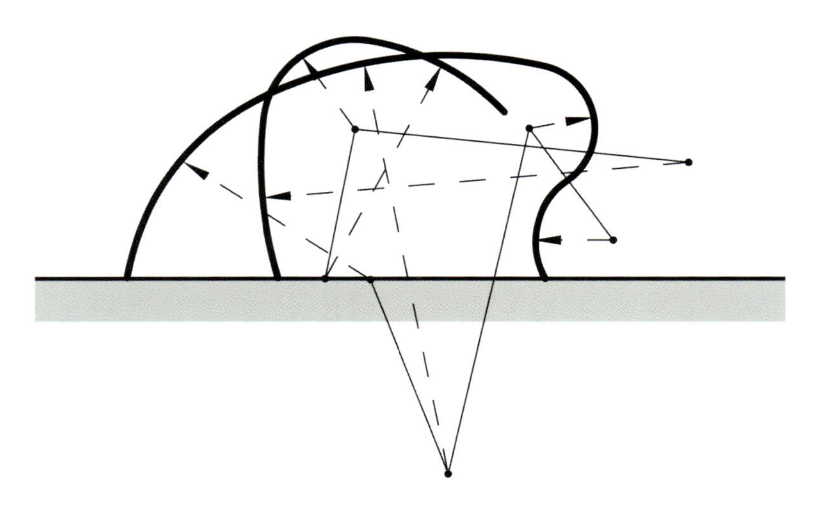

A similar approach was applied at another anonymous suburban railway station, 'Transtation O' (1997), where Endo used corrugated steel bent into three different profiles – a cantilever with one end of a sheet fixed in the ground, a gate-shaped frame attached to the cantilever type and a section in which both sides of the plate are fixed to the ground – to create a long, unbroken sequence of undulating steel ripples. The effect is discreet, stylish and entirely unexpected.

Endo has used corrugated iron for a wide variety of building types. His built projects include a research centre for a fishery cooperative, protective canopy for a used car dealer, private houses and office buildings. Endo's headquarters for a construction company in Nishinomiya is an example of his 'Roofecture' building type, in which, 'walls and roofs are replaced by continuous ribbons of sheet steel'.

However, the project for which he may be best known is a public toilet at a public park in Hyogo (completed 1998). 'Springtecture H' – the name refers to the way sheets of corrugated iron are coiled, suggesting tension – has three components: ladies and gentlemen's toilets and the park keeper's office. The need for both private and accessible facilities within the same building was perfectly suited to Endo's interest in using unbroken sheets of metal to form the outer shell of a building as well as the floors and roofing. The outcome resembles a curling spiral of apple peel, and has become a popular icon for an emerging residential district in the mountains of Hyogo.

Like all of Endo's corrugated steel buildings, the public toilets make sense to people of all ages. He has used the structural characteristics of an industrial material to create buildings that are inventive, playful and above all fun.

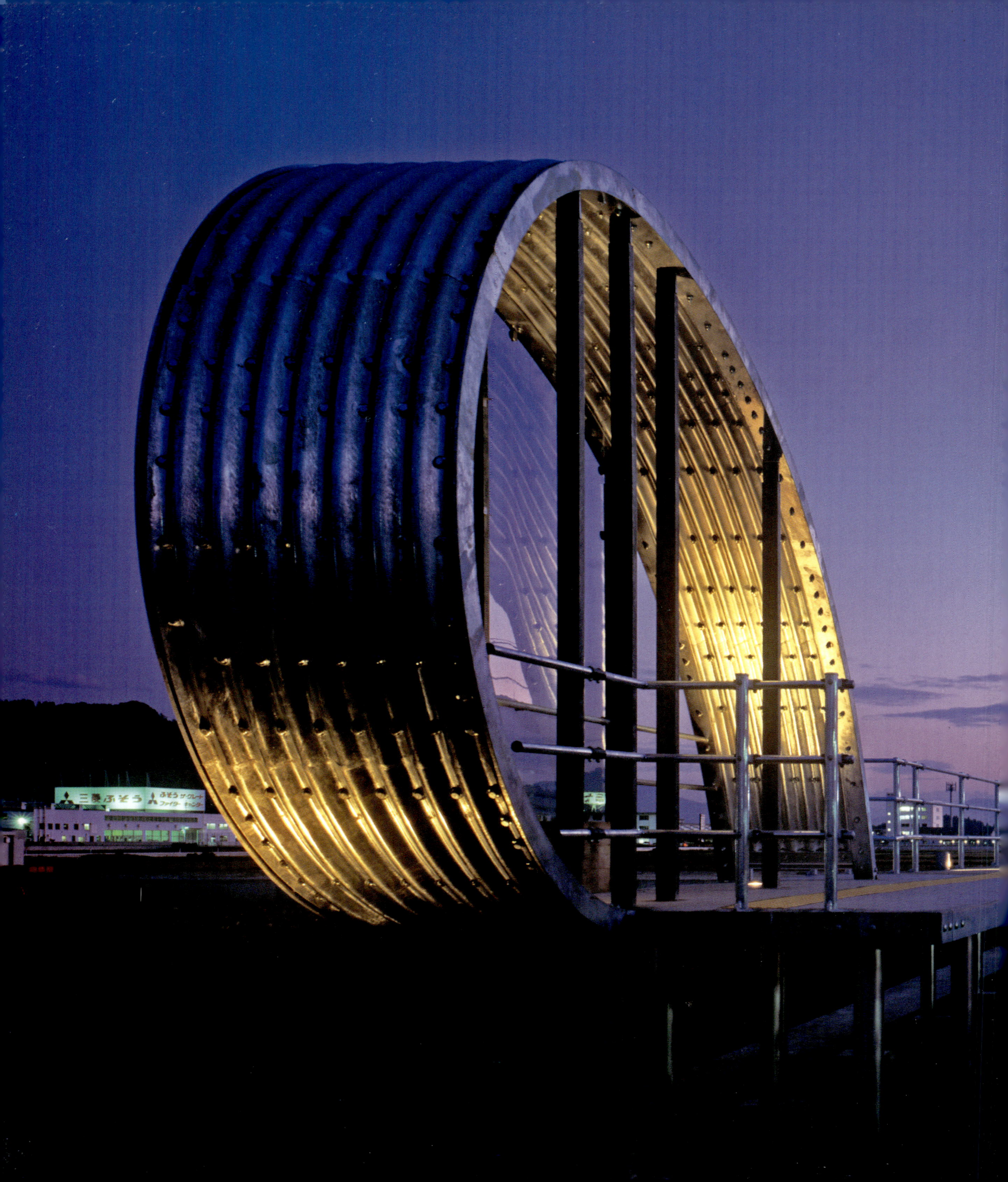
三菱ふそう

Corrugated iron bent into teardrop-shaped loops lend form and identity to a long, thin railway station at Fukui.

Three-storey office building for a construction company in Nishinomiya, an example of Endo's 'Rooftecture' building type in which walls and roofs are replaced by continuous ribbons of sheet steel'.

ABOVE Public toilets and park keeper's office, Hyogo (1998). The private and public areas are accommodated within the same unbroken coil of corrugated steel.

BELOW AND OPPOSITE The use of a continuous strip of corrugated steel means that interior walls double as exterior ceilings and floors, which in turn extend as exterior walls/roofs before turning once again into internal spaces.

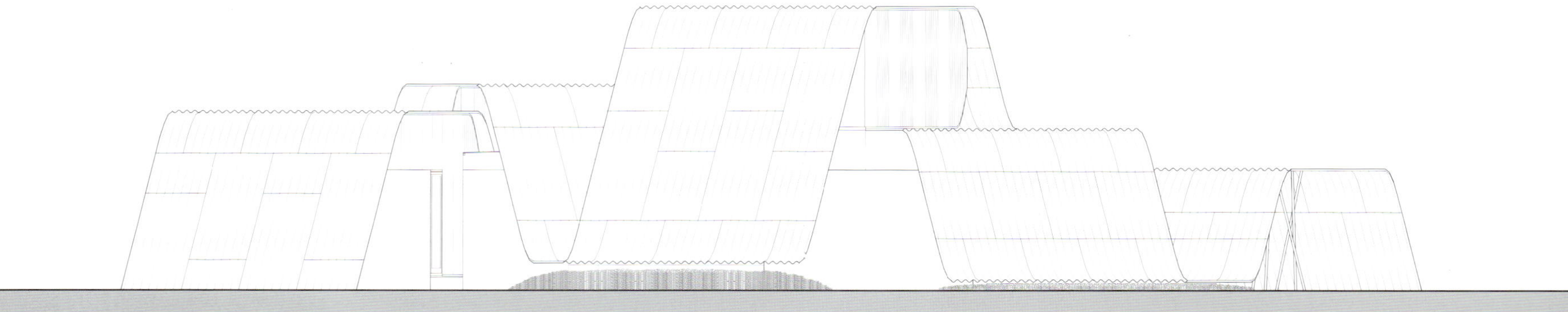

ABOVE 'Springtecture B', a live/work unit in Biwa-cho, Shiga (2002).

OPPOSITE 'Slowtecture Harada', a multi-purpose assembly hall in Kamigori-cho, Hyogo Prefecture (2004), is enclosed within a serpentine cylinder of weathered steel. Portholes illuminate the nautical-style interior.

Acknowledgements
The authors would like to acknowledge the generous and in some cases invaluable contributions of the many individuals and organizations who assisted during the research and development of, *Corrugated Iron – Building on the Frontier*, including Pedro Guedes, a Senior Lecturer in architecture at the University of Queensland, Australia.

Since the early 1970s, Guedes has been studying nineteenth-century iron construction and architecture. As part of his research he has developed an extensive original archive on the history of corrugated iron that he has made available to this publication. In addition, Guedes has made a substantial contribution to the book as writer, advisor and consultant, particularly in relation to the history of corrugated iron. He has suggested sources in the field, advised on appropriate images and recommended contacts in South America, South Africa and Australia.

We would also like to acknowledge the contributions of Peter Alsop, President of the Geelong Historical Society; Imelda Bargas of the New Zealand Historic Places Trust; Stephen Bainbridge, Offa's Dyke Books; Will Bruder, architect; Mark Cameron-Swan; Sheena Carmichael owner of 'Torreanbeg'; Chris Chiei architect; Mark Clarke; Jane Crawley and all at Frances Lincoln; Tony Cross of the Welsh Kite Trust; Mary Davies, daughter of Captain A. R. Turner of the East African Pioneers; Bob Dickinson; Diane Eddy of the Estates Office, Balmoral Estates; Tatabonko Orok Edem of the Niger Delta Congress; Shuhei Endo, architect; Tom Heneghan, Chair of Architecture at the University of Sydney; the Highland Folk Museum at Kingussie and Newtonmore; Des Key; Ironbridge Gorge Museum; David Lake of Lake|Flato Architects; Professor Miles Lewis, University of Melbourne; Philippa Lewis of the Edifice Picture Library; Dana Lockett of the Historic American Engineering Record in the US National Park Service; Phil Lynch at the Port of Tyne Authority; Brendan Marshall of Terra Culture Heritage Consultants, Melbourne; Michael Millichamp and Peter Webster of Lighthouse EClips; Jo Moore at the Museum of New Zealand; Brad Morrison of Primary Source Research in British Columbia; Larry Newitt and Claude Tremblay of the Geological Survey of Canada; Sigga and Paul Newton, owners of Hóll Cottage in Reykjavík; North of England Institute of Mining and Mechanical Engineers; Alasdair Ogilvie, photographer; Jeremy Pitts of Reed Pitts Architecture; Stephen Pomeroy; David Redfern, of the Kidderminster Museum Trust; Phil Priestly; David Redford of the Woodhall Spa Museum Trust; Jan Erik Ringstad of the Sandefjord Whaling Museum; Professor Dr Jesús Barquín Sanz of the Universidad de Granada; Ben Short; the Amoret Tanner Collection; and Lynne Wood.

Finally, thank you very much to Annabel Biles and Libby Lord, both of whom endured long evenings listening to stories about corrugated iron, but have remained enthusiastic and committed supporters throughout.

Bibliography
The following core sources are gratefully acknowledged:

Chapter 1
Evill, William, 'Description of the Iron Shed at the London Terminus of the Eastern Counties Railway' in *Minutes of the Proceedings of the Institute of Civil Engineers*, June 1844
Foster, Richard, *Birmingham New Street: The Story of a Great Station including Curzon Street*, Wild Swan Publications, 1997
Loudon, J. C, *An Encyclopaedia of Cottage, Farm, and Villa Architecture and Furniture*, Longman, Rees, Orme, Brown, Green & Longman, 1833
Robson's London Directory, 1832
Skempton, Professor A. W., 'The Boat Store, Sheerness (1858–60) and its Place in Structural History' in *Transactions of the Newcomen Society*, vol. 32, 1959
Turner, Richard, 'Description of the Iron Roof over the Railway Station, Lime-street, Liverpool' in *Minutes of the Proceedings of the Institute of Civil Engineers*, vol. 9, session 1849-1850
Chatham Docks
Evans, David, *Building the Steam Navy: Dockyards, Technology and the Creation of the Victorian Battle Fleet*, Conway Maritime Press, 2004
Paddington Station
Binding, John, *Brunel's Bristol Temple Meads: A study of the design and construction of the original railway station at Bristol Temple Meads, 1835–1965*, Oxford Publishing Company, 2001
Brindle, Steven, *Paddington Station: Its History and Architecture*, English Heritage, 2004
Rolt, L. T. C, *Isambard Kingdom Brunel*, Longman, Green, London, 1957

Chapter 2
Bellhouse, D. R., *David Bellhouse and Sons, Manchester*, Published privately, 1992
Blake, George Palmer (ed.), *The Great Exhibition: London's Crystal Palace Exposition of 1851. A facsimile of the Illustrated Catalogue of London's 1851 Crystal Palace Exposition*, Gramercy Books, 1995
Davies, James, *Galvanized Iron: Its Manufacture and Uses*, E. & F. N. Spon, 1899
Herbert, Gilbert, *Pioneers of Prefabrication: The British Contribution in the Nineteenth Century*, The John Hopkins University Press, 1978
Koeppel, Elliot, H., *The California Gold Country or Highway 49 Revisited: Being an Account of the Life & Times of the People and the Mining Camps*, Malakof & Co., 1995
Lewis, Miles *Portable Houses, 339 Coventry Street, South Melbourne: Statements of Significance*, National Trust of Australia, Victoria, 1993
Morris, William, *Architecture, Industry, and Wealth: Collected Papers by William Morris*, Longmans, Green and Co., London, 1902
Peterson, Charles E., 'Early American Prefabrication' in *Gazette des Beaux–Arts*, 6th series, vol. XXXIII, 1948
Robinson, Professor Brian, *Howden and Derwent: The Building of the Upper Dams of the Derwent Valley Water Board*, J. W. Northend Ltd, 2004
Sherwood, J. Ely, *Emigrant's Guide*, New York, February 1849
Stefánsson, Hjörleifur, *Bárujárn*, Minjavernd, Reykjavík, 1995
Tomes, Robert, *Panama in 1855, An Account of the Panama Rail-Road of the Cities of Panama and Aspinwall, With Sketches of Life and Character on the Isthmus*, Harpers and Brothers, 1855
A wide selection of catalogues and advertisements were consulted for this chapter, including examples by the following manufacturers:
Boulton & Paul, Ltd; Frederick Braby & Co.; William Cooper; Hill and Smith; Isaac Dixon & Co.; John Lysaght Ltd; The Porter Iron Roofing Co.; David Rowell & Co.; Richard Walker; Charles D. Young and Company
Brown Brothers iron store
Conservation management plan for the Former Brown Brothers' Store, prepared for Heritage Council, Victoria, Australia (Historical notes by Peter Alsop, Appendix 1) *Brown Brothers Store*, an analysis by Miles Lewis, 31 May 1989
Pilgrims Rest
Jeppe, Frederick, 'The Zoutpansberg Goldfields in South Africa', *The Geographical Journal*, vol. 2, 1893
Russell, Annie, 'A Woman in the African Diggings', *The Century – a Popular Quarterly*, vol. 45, issue 5, September 1893
Grytviken
Basberg, Bjorn., *The Shore Whaling Stations at South Georgia: A Study in Antarctic Industrial Archaeology*, Novus Forlag, 2004
Hart, Ian B., *Pesca – Compania Argentina de Pesca Sociedad Anonima of Buenos Aires: A History of the Pioneer Modern Whaling Company in the Antarctic*, Aidan Ellis, 2001

Chapter 3
Denison, Simon, *Quarry Land: Impermanent Landscapes of the Clee Hills*, Greyscale Books, 2005
Lewis, Arthur, *The Life and Work of E. J. Peck Among the Eskimos*, Hodder & Stoughton, 1904
Muthesius, Stefan, 'The 'Iron Problem' in the 1850s in Architectural History', *The Journal of the Society of Architectural Historians of Great Britain*, vol. 13, 1970
Robertson, E. G., *Victorian Heritage: Ornamental cast iron in architecture*, Melbourne, 1974
Smith, Ian, *Tin Tabernacles: Corrugated Iron Mission Halls, Churches and Chapels of Britain*, Camrose Organisation, 2004
Shrubland Road Church, Hackney
Baker, T. F. T., *Victorian History of the County of Middlesex, vol. X: Hackney Parish*, Published for the University of London Institute of Historical Research by Oxford University Press, 1995
Cox, J., *An Account of the Parish of St Leonard, Shoreditch, Middlesex* (An unpublished, handwritten document of 508 pages compiled over 20 years, collected 'from various sources')
Watson, Isobel, 'Built to last: The iron chapel, Shrubland Road' in *The Terrier, periodical of the Friends of Hackney Archive*
St John's Church, British Columbia
Healey, Edna, *Lady Unknown: The Life of Angela Burdett-Coutts*, New York, 1978
Underhill, Stuart, *The Iron Church 1860–1985*, British Columbia, 1984

Chapter 4
Captain Binney R. E. *et al*, 'Report on Hutting made by the Board of Officers assembled in the Crimea,' Professional Papers of the Corps of Royal Engineers, Vol. VII, 1858
Parkes MD, E. A., *Report on the Formation and General*

Management of Renkioi Hospital, 1857
Pawley, Martin, 'Lost Footprints in the Air', in Dean, Christopher (ed.), *Housing the Airship*, Architectural Association, London, 1986
Nissen Huts
Anon, *Memorandum concerning Nissen Huts (16' span x 36.8' long), including instructions for erection; specifications, weights, drawings and illustrations*, Nissen Buildings Ltd, Rye House, Hoddesdon, Hertfordshire, September 1939
McCosh, Fred, *Nissen of the Huts, a biography of Lt Col. Peter Nissen DSO*, B.D. Publications, United Kingdom, 1997
Quonset Huts
Decker, Julie and Chiei, Chris, *Quonset Hut: Metal Living for a Modern Age*, Princeton Architectural Press, 2005
Cardington Hangars
Chamberlain, Geoffrey, *Airships – Cardington. A history of Cardington airship station and its role in world airship development*, Terence Dalton Limited, Lavenham, 1984

Chapter 5
Anderson, Scott, *The Man Who Tried to Save the World: The Dangerous Life and Mysterious Disappearance of Fred Cuny*, Anchor Books, New York, 1999
Architecture for Humanity (ed.), *Design Like You Give a Damn – Architectural Responses to Humanitarian Crises*, Thames & Hudson, 2006
Davis, Mike, *Planet of Slums*, Verso, New York, 2006
Guedes, Pedro (ed.), *Encyclopedia of Architectural Technology*, McGraw-Hill, New York, 1979
Hatzfeld, Jean, *Machete Season – the Killers in Rwanda Speak*, Farrar, Straus and Giroux, New York, 2003
Parsons, Timothy, '"Kibra is Our Blood": The Sudanese Military Legacy in Nairobi's Kibera Location, 1902–1968,' *The International Journal of African Historical Studies*, vol. 30, number 1, 1997, pp.87-122

Chapter 6
Amery, Colin, *Architecture, Industry and Innovation: Work of Nicholas Grimshaw and Partners, 1965-88 vol. II*, Phaidon, London, 1995
Buisson, Ethel and Billard, Thomas, *The Presence of the Case Study Houses*, Birkhäuser, Basel, 2004
Goad, Philip and Bingham-Hall, Patrick, *New Directions in Australian Architecture*, Pesaro Publishing, Australia, 2001
Goad, Philip, *Troppo – Architecture for the Top End*, Pesaro Publishing, Australia, 2005
Jenkins, David (ed.), *Norman Foster Works I*, Prestel, 2002
Peter Myers, Peter, 'Corrugated galvanised iron: the profile of a national culture' in *Transition*, June 1981
Ogg, Alan, *Architecture in Steel, the Australian context*, The Royal Australian Institute of Architects, 1987
Rosa, Joseph, *Albert Frey, Architect*, Rizzoli, New York, 1990
Smith, Elizabeth A. T., *The Complete CSH Program 1945–1966*, Taschen, Cologne
Glenn Murcutt
Drew, Philip, *Leaves of Iron – Glenn Murcutt: Pioneer of an Australian Architectural Form*, The Law Book Company Ltd, Australia, 1985
Fromonot, Francoise, *Glenn Murcutt: Buildings and Projects, 1962–2003*, Thames and Hudson, 2005
Lake|Flato Architects
Ojeda, Oscar Riera (ed.), *Lake|Flato: Buildings and Landscapes*, Rockport Publishers, 2005
Shuhei Endo
Hiroyuki Suzuki (ed.), *Paramodern Architecture: Shuhei Endo*, Electa Architecture, 2002

The following are among the journals, newspapers and periodicals consulted for this book:
Architectural Review; The Civil Engineer and Architect's Journal; The British Architect; Building News; The Builder; The Engineer; The Journal of the Society of Architectural Historians; Illustrated London News; Life Magazine; Lloyds Newspaper; Mechanics Magazine; The Melbourne Age; Practical Mechanics Journal; Sydney Morning Herald; The Times; Transition

Picture sources and credits
Numbers refer to page numbers

While every effort has been made to trace all copyright holders, the authors would like to apologize in advance for any that may have been inadvertently overlooked.

Chapter 1
P2–3: Alasdair Ogilvie **P4–5**: Peter Bennetts, courtesy Jesse Judd **P6**: Rob Duker/Noero Wolff Architects **P7**: AP/EMPICS **P8/9**: Simon Holloway **P10**: British patent number 5786: Henry Robinson Palmer, 28 April 1829 **P11**: J. C. Loudon, *Cottage, Farm, and Villa Architecture and Furniture...* (Frederick Warne & Co., London, 1846) p.207 (University of Queensland Library [UQL]) **P12** (top): Image originally published in, *The Lysaght Century 1857-1957*, John Lysaght Limited, St Vincent's Works, Bristol, 1957. With thanks to Bluescope Lysaght **P12** (bottom): British patent number 10399: John Spencer, 23 November 1844 **P13**: Originally published in *Robson's Commercial Directory of the Six Counties forming the Norfolk Circuit...*, William Robson & Co., London, c.1839 **P14** (top): *Illustrated London News (ILN)*, vol.17, 1850, p.477 (UQL) **P14** (middle): *ILN*, vol.15, 1849, p.380 (UQL) **P14** (bottom): *ILN*, vol.25, 1854, p.184 (UQL) **P15**: Edifice **P16**: *ILN*, vol.54, 1869, p.421 (UQL) **P17**: *ILN*, vol.22, 1853, front cover (Simon Holloway private collection) **P18**: Edifice **P19** (top): *ILN*, vol.24, 1854, p.505 (UQL) **P19** (bottom): *Minutes of the Proceedings of the Institute of Civil Engineers*, vol.9, 1850, Plate 10 (UQL) **P20**: Supplied by Kidderminster Museum Trust (Simon Holloway private collection) **P21/22**: Simon Holloway **P23** (top): Edifice **P23** (bottom): Simon Holloway **P24**: *The Builder*, vol.12, 1854, p.291 (UQL) **P25**: Supplied by Kidderminster Museum Trust (Simon Holloway private collection) **P26**: Simon Holloway **P27**: Ewing, Matheson, *Works in Iron: bridge and roof structures*, E. & F. N. Spon, 1877, 2nd ed., London, p.217. Courtesy of Pedro Guedes

Chapter 2
P28: National Library of Australia, nla.pican8441322 **P30**: *ILN*, vol.16, 1850, p.5 (UQL) **P31** (top): *ILN*, vol.14, 1849, p.109 (UQL) **P31** (middle): *ILN*, vol.15, 1849, p.20 (UQL) **P31** (bottom): All pictures from an article titled, 'Iron as a Building Material' published in the *Practical Mechanics Journal*, vol. 5, 1852, pp.273-75 **P32**: Special Collection, Baillieu Library, University of Melbourne **P33** (top): National Library of Australia, nla.pican8713100 **P33** (bottom): RIBA Library Photographs Collection **P34** (top): Edifice **P34** (bottom): British Columbia Archives, G-01986 **P35**: *ILN*, vol.19, 1851, p.613 (UQL) **P36**: *ILN*, 27 June 1857, p.635 **P36**: *The Engineer*, vol.1, 1856, p.244 (UQL)37: *The Expositor*, vol. for 1851, p.233 **P38** (top): *ILN*, vol.24, 1854, p.48 (UQL) **P38** (middle): *The Builder*, vol.11, 1853, p.422 **P38** (bottom): Simon Holloway **P39** (top): Simon Holloway private collection **P39** (bottom), **P40** and **P41** (top): Alasdair Ogilvie **P41** (bottom left): *ILN*, vol.56, 1870, p.257 (UQL) **P41** (bottom right): *ILN*, vol.30, 1857, p.190 (UQL) **P42** (top): *ILN*, vol.62, 1873, p.488 (UQL) **P42** (bottom): Supplied by the Peak District National Park Authority **P43**: *Building News*, vol.45, 1883, p.385 (UQL) **P44**: Edifice **P45** (both illustrations): From Isaac Dixon & Co. catalogue 'Of Improved Iron Buildings for all purposes,' c.1885 (Simon Holloway private collection) **P46** (top): Edifice **P46** (bottom): Excerpt from John Lysaght Ltd catalogue, 1910 (Simon Holloway private collection) **P47**: From Boulton and Paul catalogue c.1900 **P48/49**: Alasdair Ogilvie **P50** (top): Simon Holloway **P50** (middle): From Boulton and Paul catalogue c.1900 **P50** (bottom): Alasdair Ogilvie **P51** (top): Simon Holloway **P51** (bottom): Excerpt from Hill and Smith Ltd advertisement c.1900 (Simon Holloway private collection) **P52/53**: Excerpt from Hill and Smith Ltd's List No.349, c.1900 (Simon Holloway private collection) **P54/55**: Pictures supplied by Brendan Marshall of Terra Culture, Melbourne **P56/57**: Adam Mornement **P58** (top): Pedro Guedes (private collection) **P58** (bottom): *Practical Mechanics Journal*, vol. for 1854, plate 154 (UQL) **P59**: Pedro Guedes private collection **P60** (top): *Civil Engineer and Architect's Journal*, 1854, plate 20 **P60** (bottom): Pedro Guedes private collection **P61** (left): *The Engineer*, 18 March 1870, p.158 **P61** (right): Supplied by Oriana Abbondanza of Gonzalez Byass **P62/63**: Simon Holloway **P64**: *ILN*, vol.16, 1850, p.229 (UQL) **P65/66** (all images): Pedro Guedes **P67**: *The Builder*, vol.25, 1867, p.391 (UQL) **P68** (top): *ILN* vol.53, 1868, p.364 (UQL) **P68** (bottom): State Library of Queensland image no. neg 14026 **P69** (top) Frederick Braby & Co. (Limited) catalogue, 1869 **P69** (bottom): State Library of Queensland image no. neg 22807 **P70/71**: State Library of Queensland image no. neg 91664 **P72/73**: Southampton Oceanographic Centre **P74–77**: All images by Roderick Eime

Chapter 3
P78/79: Alasdair Ogilvie **P80** (top): *ILN*, vol.5, 1844, p.138 (UQL) **P80** (bottom): *ILN*, vol.5, 1844, p.208 (UQL) **P81** (top): National Library of Australia, nla.pic-an8713118 **P81** (bottom): *ILN*, vol.22, p.324 (UQL) **P82** (top): *The Builder*, 7 April 1883 (UQL) **P82** (bottom): Supplied by Claude Tremblay, Centre d'Etudes Nordiques and Larry Newitt at NRCan **P83** (top): *The Builder*, vol.13, 1853, p.507 (UQL) **P83** (bottom): *The Builder*, vol.13, 1853, p.507 (UQL) **P84–86**: All images by Alasdair Ogilvie **P87** (both images): *Instrumenta Ecclesiastica*, vol. 2, London, 1856 (Pedro Guedes private collection) **P88/89**: Simon Holloway private collection **P90**: *Tee-Square and Tape – a quarterly Magazine for Architects* issued by David Rowell & Co., vol.3, 1909, p.48 (Pedro Guedes private collection) **P91**: Metropolitan Borough of Hackney Archives, HA10399 **P92/93**: Simon Holloway **P94** (both pictures): Methodist and Congregational Archives, Portsmouth City Records Office **P95**: Alasdair Ogilvie **P96–98**: Pedro Guedes **P99**: Alasdair Ogilvie **P100**: *Building News*, 22 May 1857 **P101**: British Columbia Archives, A-02788 **P102**: British Columbia Archives, C-03847 **P103**: British Columbia Archives, B-07346 **P104/105**: Doug Houghton Photography

Chapter 4
P106: Australian War Memorial, OG0179. Photographer, John T. Harrison **P107**: Australian War Memorial, P00579.004 **P108**: Florence Nightingale Museum Trust, London **P109** (top): RIBA Library Photographs Collection **P109** (middle): *ILN*, vol.27, 1855, p.397 (UQL) **P109** (bottom): *ILN*, vol.31, 1857, p.324 (UQL) **P110**: Museum of New Zealand Te Papa Tongarewa, C.003368 **P111** (top): *Mechanics' Magazine*, 30 August 1856 **P111** (bottom): Wilson H. W., *After Pretoria: The Guerrilla War*, The Amalgamated Press, London, *c*.1901 **P112** (top): From Isaac Dixon & Co. catalogue 'Of Improved Iron Buildings for all purposes,' *c*.1885 (Simon Holloway private collection) **P112** (bottom): Imperial War Museum, 40791 **P113**: Imperial War Museum, HU64203 and HU64201, with thanks to Mrs Mary Davies, daughter of Captain A. R. Turner, MC. **P114**: Luke Hamilton Larmuth, *Airship Sheds and their erection, Minutes of the Proceedings of the Institute of Civil Engineers*, plate 9, vol. 212, session 1920-21, Part II **P115**: Australian War Memorial, E04150 **P116**: Imperial War Museum, D778 **P117** (top left): Imperial War Museum, D5949 **P117** (top right): Imperial War Museum, D2182 **P117** (bottom): Imperial War Museum, D3606 **P118/119**: Imperial War Museum, Q2529 **P120** (top left): Imperial War Museum H40790 **P120** (top right) Imperial War Museum H40793 **P120** (bottom): Imperial War Museum, Q1402 **P121** (top): Imperial War Museum, Q2783 **P121** (bottom): Imperial War Museum, Q1711 **P122**: Edifice **P123**: Caroline Mornement **P124**: Copyright, Ted Kane, www.polarinertia.com **P125** (top): Library of Congress, Prints and Photographs Division, Historic American Buildings Survey, Reproduction Number, HABS HI,2-EWA.V,1E-3 **P125** (bottom): Library of Congress, Prints and Photographs Division, Historic American Buildings Survey, Reproduction Number, HABS HI,2-EWA.V,1B-6 **P126/127**: Library of Congress, Prints and Photographs Division, Historic American Buildings Survey, Reproduction Number, HABS AK-34-JJ. Photographer, Jet Lowe **P128–129** (all images): Copyright, Ted Kane, www.polarinertia.com **P130** (top): Getty Images (3422401) **P130** (bottom): Getty Images (3319109) **P131** (top and bottom): Simon Holloway

Chapter 5
P132: Magnum Photos (LON15787). Photographer, Ian Berry **P133**: From *Lysaght Referee*, 31st edition, produced by Lisa J. Carrick of Bluescope Steel Limited, pp.78-84 **P135** (both images): Hannah Mornement **P136/137**: David Southwood/Noero Wolff Architects **P138**: Magnum (LON42902). Photographer, Stuart Franklin **P139** (all images): Hannah Mornement **P140**: Getty Images (56576061). Photographer, Eric Feferberg **P140**: Getty Images (56283755). Photographer, Banaras Khan **P142** (two images on left): Supplied by the Swiss Federal Institute for Aquatic Science and Technology (Eawag) **P142** (two images on right): Adam Mornement **P143**: Hannah Mornement

Chapter 6
P145: J. Paul Getty Trust. Used with permission. Julius Shulman Photography Archive Research Library at the Getty Research Institute, Los Angeles **P145**: The Estate of R. Buckminster Fuller **P146/147**: J. Paul Getty Trust. Used with permission. Julius Shulman Photography Archive Research Library at the Getty Research Institute, Los Angeles **P149**: Getty Images (50874845). **P150–153**: Bill Timmerman/Will Bruder Architects **P154/155**: Bill Timmerman/blank studio **P156** (both images): Urs Peter Flueckiger **P157** (both images): Chipper Hatter **P158**: Eric Sierins **P160**: National Library of Australia, nla.pic-an10642263-27 **P161** (all images): Troppo Architects **P162/163**: Patrick Bingham-Hall, Pesaro Publishing **P164** (top): Bowali Visitors Information Centre/Troppo Architects **P164** (bottom): Adam Mornement **P165–167**: Peter Bennetts, courtesy Jesse Judd **P168** (both images): Norman Foster **P169** (both images): Tim Street-Porter **P170** (both images): Tessa Traeger **P171/172**: John Donat/RIBA Library Photographs Collection **P173** (both images): Jo Reid and John Peck/Grimshaw **P174/175**: Werner Huthmacher, Berlin **P176/177**: Hans Werlemann (Hectic Pictures) **P178/179**: Christian Richters **P180–191**: All images by Max Dupain **P192–203**: All images by Paul Hester, except pp.196 and 197, by Dawn Jones, and p.202, courtesy of Lake|Flato Architects **P204–217**: All images by Yoshiharu Matsumura, except p.204 and pp.208-9, by Toshiharu Kitajima **P224** John Lysaght advertisement (Simon Holloway private collection)

INDEX

Numbers in *italics* refer to illustrations

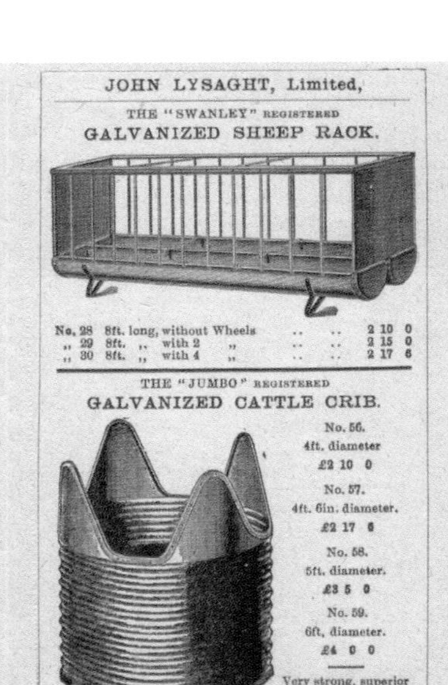

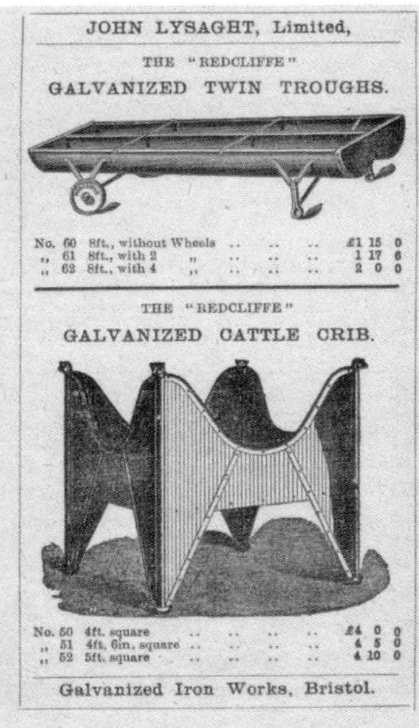

These Sheds form the most secure method of Storing Crops in sound condition. They can be erected of any size.

We shall be pleased to quote prices on receipt of particulars for any description of Building, fixed complete in any part of the kingdom.

IRON BUILDINGS FOR AGRICULTURAL AND MANUFACTURING PURPOSES.

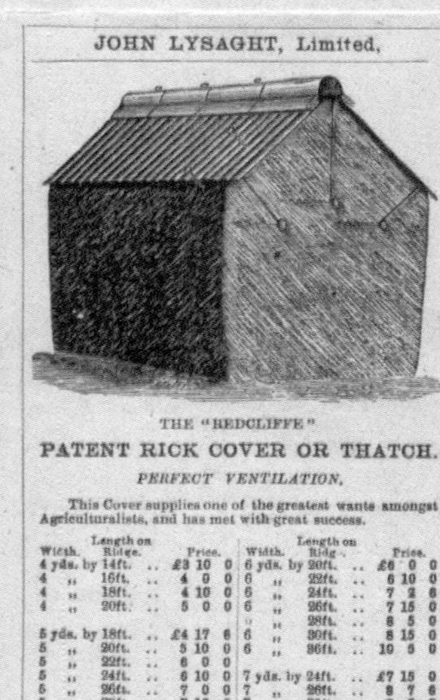

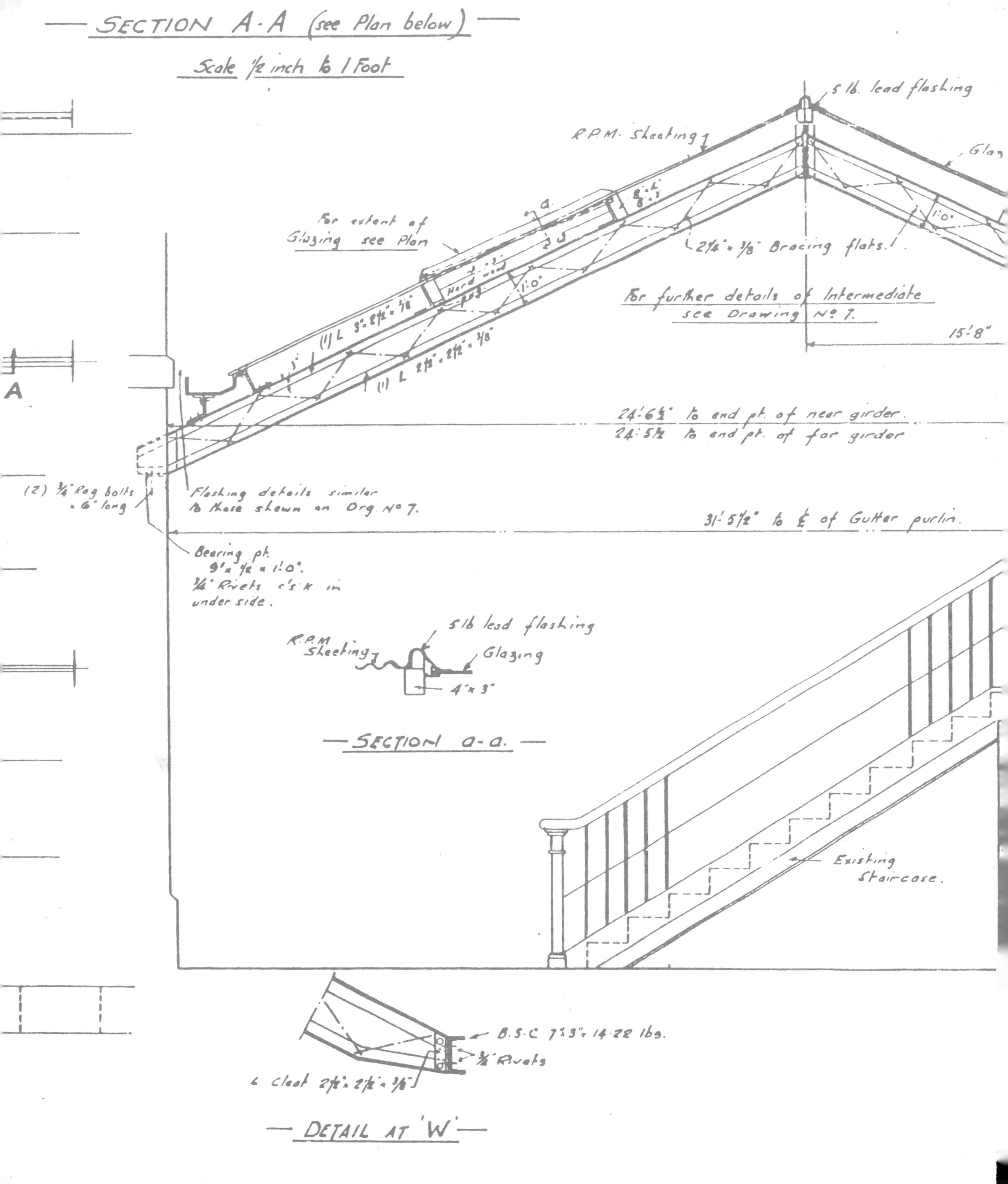
SECTION A·A (see Plan below)
Scale ½ inch to 1 Foot
5 lb. lead flashing
R.P.M. Sheeting
For extent of Glazing see Plan
2¼" × ⅜" Bracing flats.
For further details of Intermediate see Drawing Nº 7.
15'-8"
24'-6½" to end pt. of near girder.
24'-5½" to end pt. of far girder
(2) ¾" Rag bolts × 6" long
Flashing details similar to those shown on Drg. Nº 7.
31'-5½" to ℄ of Gutter purlin.
Bearing pt. 9" × ½" × 1'-0". ¾" Rivets c's'k in under side.
5 lb lead flashing
R.P.M. Sheeting
Glazing
4" × 3"
SECTION a-a.
Existing Staircase.
B.S.C 7" × 3" × 14·22 lbs.
⅞" Rivets
L Cleat 2½" × 2½" × ⅜"
DETAIL AT 'W'